Correlational Procedures for Research

# CORRELATIONAL PROCEDURES FOR RESEARCH

*Robert M. Thorndike*

**WESTERN WASHINGTON UNIVERSITY**

GARDNER PRESS, INC., NEW YORK

Distributed by Halsted Press
A Division of John Wiley & Sons, Inc.

New York  •  Toronto  •  London  •  Sydney

Gardner Press, Inc.
19 Union Square West
New York 10003

Distributed solely by the Halsted Press Division
of John Wiley & Sons, Inc., New York

Library of Congress Cataloging in Publication Data

Thorndike, Robert M.
Correlational procedures for research.

"A Halsted Press Book."
1. Correlation (Statistics)  I. Title.
QA278.2.T48   1976      519.5'37      76-8462
ISBN 0-470-15090-4

Printed in the United States of America

# PREFACE

Developments in computer technology and the rapid spread of large, high-speed computing equipment in the 1950s and early 1960s has led to a new kind of book on multivariate analysis, a book designed to meet the needs of a new group of users of multivariate techniques. Until 1962 the few books on multivariate analysis were written for the mathematically sophisticated. In that year Cooley and Lohnes published the first book aimed at the researcher in the field rather than the student of statistics. Their book, *Multivariate Procedures for the Behavioral Sciences*, was timely because it showed researchers with multivariate problems how to bring the power of the computer to bear on research issues requiring multivariate answers. Everyone had a computer; the question was how to use it and what to use it for.

The years since 1962 have seen an ever-increasing number of books written for users of multivariate analysis. This period has also seen a marked increase in the number of people with a new function on the research team, the statistical consultant or data analyst. These people, of whom the present author is one, work at the interface between the researcher and his computer. The data analyst is the one called on to suggest a method of analysis and to provide a 15 minute to 2 hour description of the analysis alternatives and consequences.

It is typical for the analyst-consultant to find that his "clients" have a fairly good background in univariate analysis of variance. Most graduate programs provide a reasonably comprehensive course in this aspect of statistics. Correlational procedures, on the other hand, seem to be given short schrift in our graduate education. Therefore, it has often been necessary to spend considerable time on basic correlation before getting to the real problem at hand, be it multiple regression or factor analysis.

One solution to the problem is to suggest some reading, and I have on numerous occasions lent copies of books that I felt did a good job of presenting the fundamentals of the desired procedure. The typical result has been that the book was unintelligible, and the reason seems to be twofold. First, there is insufficient understanding of the very basic concepts of correlation, and second, the books contained too many matrix equations and derivations and not enough verbal explanation.

Desperation is the mother of invention, and this book is the result of my desperation. It has been designed and written to fill the gap between no knowledge and the level of sophistication about correlation that is assumed by the other books available on multivariate analysis and to provide me (and others like me) with something my "clients" can and will read. The approach is largely intuitive and geometric rather than mathematical, and it follows my own, largely intuitive and nonmathematical way of developing an understanding of what are, at the same time, quite simple and highly complex concepts.

When one's understanding of complex relationships is based on intuition and experience rather than on formal mathematics, it is very difficult to communicate to others the insights one has developed. In the absence of mathematical training they must develop their own insights and their own intuitive feeling. I have taught multivariate statistics for 4 years using the format developed in this book with more success than I think I could have achieved by a conventional approach. Therefore, this book is also written as a primary text for courses like mine or as supplementary reading for students in more mathematically oriented multivariate statistics courses.

*Correlational Procedures* is not intended to be a stand-alone reference on multivariate statistics. The final equations for most of the concepts discussed are given, but the reader is asked either to accept them on faith or to consult the sources cited in which the derivations are provided. Rather than being a definitive mathematical treatment, this is an overview and introduction. The work of other authors, notably Quinn McNemar, Harry Harman, and Bill Cooley and Paul Lohnes, is frequently cited for the reader who wants or needs the mathematical foundations of the topics discussed. There is a conscious effort in this book to avoid mathematical detail and to focus instead on the concepts involved and on research methodology. At times the discussion will warn the reader of mistakes that are commonly made and describe what I feel are the best ways to do things. The emphasis is always on conceptual understanding, the conduct of research, and the interpretation of results, rather than on the how and why of the equations.

The organization of the book reflects a conviction that understanding can be most readily developed by showing the essential unity and orderly progression of concepts in multivariate statistics. Once a few basic ideas, which can be expressed in relatively simple cases, are thoroughly understood, the remaining concepts are shown to be direct generalizations. The geometric interpretation of the correlation coefficient as the angle between two vectors in a people space is readily generalizable to multiple and canonical correlation. The concept of composites of variables, which is developed in the discussion of multiple correlation, applies directly to all of the subsequent procedures.

Most of the concepts of multivariate procedures are introduced in Part I, which considers the case of relationships between two variables. Part II deals with those techniques in which single variables are replaced by sets of variables. For example, a bivariate correlation problem becomes a multiple correlation problem when one of the single variables is replaced by a set of variables. When both of the single variables are replaced by sets of variables, the result is a canonical correlation. The term *external factor analysis* is used to refer to the procedures described in Part II. Part III considers the case in which there is only one set of variables, and the techniques described are called *internal factor analyses*. I feel that these terms, coined by M. S. Bartlett, accurately reflect the similarities and differences of the procedures described in the book.

Any book, whether it has one author or ten, is really the work of many people. There are several whose aid has been invaluable in the preparation of this book. First among them is Miss Joy Dabney who prepared the figures. She spent countless hours making sense out of my rough sketches and converting them into comprehensible diagrams. To her go my special thanks. Sincere thanks also go to Mrs. Jane Clark and her staff at the Bureau for Faculty Research at Western Washington State College for their careful typing of the several drafts of the manuscript. And finally, to my students, who have helped shape this book, who have worked patiently with successive draft versions and who have helped me to clarify my own thinking about the topics it contains, my thanks.

*Robert M. Thorndike*

# CONTENTS

# PART I

# BIVARIATE RELATIONSHIPS

# 1

# THE NATURE OF DATA

The business of science is the collection and organization of information concerning the world around us with the goal of increasing our understanding of the things we observe. Throughout the history of science there is a consistent pattern related to increasing our knowledge. The first step is always observation of phenomena and the recording of those observations. This is followed by attempts to organize the observations in some way so that we may, given a new observation of a phenomenon we have observed before, predict some other aspects of the phenomenon. The final level of the scientific endeavor is the development of explanations of the causes of the things we have observed. These explanations are then tested by performing experiments in which we control certain aspects of the situation to rule out some explanations, make predictions of what our observations will be, and then test the accuracy of our predictions.

A classic example of science at work is the development of the law of gravity. For centuries people had observed various regularities in the world around them: things fell from high places to lower places most

**3**

of the time, there was a general regularity to the motion of stars, and so forth. At this pre-Newtonian stage classification and organization of observations had taken place, and some predictions could be made about what would happen if one held an object over one's head and then let go of it. However, quantitative predictions were not very accurate. Then the law of gravity was proposed to explain the behavior of falling objects. Experiments were performed that ruled out competing explanations and refined the law to the point where very accurate predictions could be made about how an object would behave under given conditions. The law was extended to explain the behavior of celestial bodies, military projectiles, and a variety of other phenomena and has become so well understood and so efficient that it now functions as a dependable technological tool, not as a subject of scientific investigation.

Most areas of study in the behavioral sciences are at the stage where the study of falling objects was when Newton sat under the apple tree. We have made many observations and attempted to classify and organize them, but we do not yet have much in the way of formal statements about these observations which can be used to make highly accurate predictions. A large part of the difficulty behavioral scientists have encountered stems from the quality of our observations and the properties we can attribute to our records of them. We do not have much to say concerning measurement in the behavioral sciences in this book. However, there are a few measurement problems with which we must deal, and now is as good a time as any.

When we make observations of phenomena, we generally try to quantify these observations in some way; that is, we wish to assign numbers to our observations according to a set of rules. The property of the objects to which we are attending is called a variable (assuming that different individual objects in the class of objects can have different amounts of the property). The property is also called a trait or an attribute, but because we are only concerned with situations in which individuals vary with respect to amount of the trait, the term *variable* is used. For example, if we are attending to the property "weight" in a group of children, weight is a variable if not all children have the same value of this property. If all the children receive the same value for weight, then weight is not a variable, because the values recorded from our observations do not vary. The recorded result of an observation (measurement) of a variable property (trait) is called a datum, and a group of these records constitutes the *data* we may try to organize and describe, using the procedures discussed in later chapters of this book.

## SCALES OF MEASUREMENT

The data we get from our observations of the standing of objects on the variable of interest are, generally, in the form of numbers. If they are not, then we must convert them into numbers before we can use statistical procedures to describe them. This would be the case, for example, if we were making observations of the activity level of children. We might record our observations as verbal reports of how active each child was. Before we could apply any statistical analysis, it would be necessary for us to assign a number to each child to represent his level of activity. We might decide to break our group of children down into three groups, those who were described as very active, those who showed a moderate activity level, and those who showed little activity, and assign the number 3 to members of the first group, the number 2 to members of the second, and 1 to members of the third. Or, we might rank order all the children and assign to each the number of his rank. In either case, our data would then be in numerical form, as required for the application of statistical procedures.

When we assign numbers to our observations, we give those numbers additional meaning. Numbers by themselves are very easy to work with and to understand because we have dealt with them as representations of quantities of things; for example, I have three apples. When used in this way, we all know what the numbers mean and we know that if I have three apples and you have six apples, then you have twice as many apples as I have. When numbers are used in this way, we refer to them as cardinal numbers. However, the assignment of additional meaning to the numbers when they are used to represent quantitative amounts of a trait (i.e., as values of a variable) makes the matter more complex. We know that 2 is twice as much as 1 and that the difference between 1 and 2 is the same as the difference between 2 and 3. But, do we really know this to be true? In our example of activity level in children, do we know that a child who received a value (score) of 2 is twice as active as a child who received a 1? Do we know that the difference between group 1 and group 2 is of the same magnitude as the difference between group 2 and group 3? The answer is that we do not know either of these things. When we add meaning to numbers by using them to represent values of a variable, we may lose some of the quantitative meaning of the numbers. When working with measurements of behavior it is important that we know what quantitative meaning our numbers have, because the kinds of statements we can make about our data and the kinds of statistical procedures we can use depend on the quantitative properties of our measurements.

The study of the quantitative properties of numbers that result from measurement is called *scaling* and is generally beyond the scope of this book. We discuss only the most fundamental concepts in this field; the interested reader should consult books such as Coombs (1964), Guilford (1954), and Torgerson (1958) for a more complete treatment of the topic.

We have a scale whenever we adopt a set of rules for assigning numbers to objects or individuals to signify their status on some variable (e.g., 3 = very active, 2 = moderately active, and 1 = inactive). A useful way of classifying measurement scales in terms of the amount of quantitative information they contain has been given by Stevens (1946). If the trait being measured is not quantitative so that the numerical values are used only to indicate group membership, the result is said to be a *nominal scale*. This is the situation that prevails when we use the number 1 to refer to boys and 2 to refer to girls. The numbers signify group membership, but they do not refer to *amounts* of anything. The same situation prevails if we use numbers to designate different schools or classrooms in a study of educational achievement. Nominal measurement is the lowest form of measurement (some experts argue that the assignment of numbers to indicate group membership does not constitute measurement at all).

The next level of measurement, which is probably the most common level attained in the behavioral sciences, results in what is called an *ordinal scale*. At this level of measurement our numbers contain information about group membership, but group membership has quantitative implications. When larger numbers are applied to a group it generally means that the members of that group have more of the trait in question than do members of groups which receive lower numbers. This is the case in our example of activity levels in children; members of the more active group were assigned a higher number than members of the less active group. The same level of measurement would hold if we ranked the children according to activity level (with 1 being lowest) and assigned numerical values on the basis of ranks. However, the numbers that form our scale still do not tell us whether the difference between group 1 and group 2 is the same as the difference between group 2 and group 3 in terms of the actual behavior the numbers are supposed to reflect. An ordinal scale conveys information about the order of groups or individuals on the trait in question, but it does not give us information about the relative magnitudes of the distances between them.

When equal distances on the numerical scale imply equal differences in quantity of the trait, we say that we have an *interval scale*. This means that equal numerical intervals refer to equal increments of the trait. Relatively pure and straightforward examples of this level of measurement are somewhat hard to find in the behavioral sciences, so we will revert to the calendar for our first example. The Gregorian calendar is an interval scale of time. The amount of time separating two events occurring at midnight, July 15 in 1602 and in 1740 is the same as the amount of time between two other events occurring at 10:15 a.m., April 4 in 566 and in 704. On the calendar, each unit of time is equal to every other unit (leap years excepted). Another, more frequently used example of an interval scale is the Celsius (centigrade) temperature scale. Here, too, relative amounts of the trait being measured are given by the numerical values of the scale. However, there is a characteristic common to both of these measuring instruments that makes them different from the situation we had when we were using numbers to represent how many apples we had. That characteristic is that for each of these interval scales the location of the point designated zero does not correspond to the zero value of the trait. Interval scales can take on negative values. There are years before the year 0 and there are temperatures below the freezing point of water.

This lack of a rationally defined zero point indicating the exact zero value of the trait is an important factor to keep in mind when interpreting the results of a statistical analysis of data collected on a scale with interval scaling properties. The reason is that the meaning attached to high and low numbers may be arbitrary, and the way that the decision is made must be kept in mind when interpreting the data. Two examples will help to make the point. First, in our study of the activity level of children (the directionality problem is just as great for ordinal scales) we assigned a value of 3 to represent membership in the most active group and a

1 to represent membership in the least active, but we might just as well have assigned these numbers in the opposite order. As a second example we may look at intelligence measurement (which has been taken by some to represent interval measurement). What would have happened if Stearn had defined the Intelligence Quotient as chronological age divided by mental age ($\times 100$)? The answer is that nothing would be importantly affected. We would have to revise some of the particulars about the way we view intelligence (for example, IQs below 100 would signify above average intelligence), but we could study mental abilities just as well with the scale reversed. It might be that our norm tables would report number of items missed rather than number correct as the raw score, but once we got used to the new direction of the scale, we could operate as efficiently as we now do.

There is a serious potential problem concerning directionality. Picture what might happen if we ranked the children from our first example on activity level and assigned rank 1 to the most active child (with successively lower levels of activity being indicated by higher numbers). Assume that we also have data on destructiveness given by the number of objects broken by each child during the year. We would expect that more active children would break more objects than would less active children (a positive relationship). To test this expectation we compute the correlation between activity and destructiveness as described in Chapter 2. Imagine our surprise when the index of relationship is found to be negative! Imagine, also, our relief when we remember that low numbers were assigned to the more active children and that we should have looked for a negative value for our index as confirmation of our expectation. Awareness of the directionality of the scale we are using is vital to sound interpretations of our results. I have seen more than one graduate student stare in bewilderment at the negative results he obtained from a correlational analysis, forgetting which end of the scale was up.

When a measurement scale has equal intervals and a nonarbitrary zero point, one which defines exactly none of the trait in question then we say that we have a *ratio* scale. This is the type of scale we have when we measure length, weight, temperature (on a Kelvin scale), or the running time of a rat in a maze. Ratio scale measurement does not result in negative values and receives its name because it is possible to make statements such as "A is twice B." This type of statement is not appropriate for an interval scale, because the zero point is arbitrarily defined. For example, it would not be reasonable to say that 60°C is twice as hot as 30°C or that a person with an IQ of 100 is twice as smart as a person with an IQ of 50 (two mentally retarded individuals together do not have the same level of ability as one normal individual, either). However, with a ratio scale such statements are meaningful. Two 100-pound individuals, together, do weigh the same amount as one 200-pound individual, and a rat that took 20 seconds to run a maze took twice as long to do it as a rat that ran the same maze in 10 seconds.

The difference between interval and ratio scales of measurement is not particularly important for most applications of correlational procedures for two reasons. First, the directionality problem is not overcome by using ratio measurement. If we are studying maze learning in the rat, for example, and we have a measure of motivation (hours of deprivation) and speed of learning (number of incorrect choices) we have two ratio scales with opposite directionality; that is, higher motivation will be represented by larger numbers and higher speed of learning will be represented by smaller numbers. Under these circumstances we would find a positive relationship between motivation and learning to result in a negative value for the correlation coefficient. The second reason why it makes little difference whether we have interval or ratio measurement when using correlational procedures is that the variables are treated as though they were measured on an interval scale by most correlational statistics. That is, in the process of calculating the correlation coefficient, which is usually the first step in applying other procedures, the variables are given arbitrary zero values equal to their means and arbitrary standard deviations equal to one. This standardization of the variables results in the conversion of all ratio scales to interval scales; thus the initial scaling properties of the measurements are irrelevant so long as they are at least at the interval level. Note that special procedures are required if the variables have been measured on a nominal or an ordinal scale.

**Accuracy of Measurement**

In addition to being aware of the restrictions different types of scales impose on his data, it is important that the investigator consider the accuracy with which his numbers reflect the true state of affairs. Almost all measurements are subject to some error in that they do not give as exact a reflection of the world as we might wish. This is a problem with which we must learn to live, but we must also be aware of its existence and implications.

Consider again the two examples of apples and active children. When I say that I have six apples, you may with reason conclude that my statement is accurate, that if there were six people in the room I would be able to provide each one with exactly one whole apple. The number is an accurate reflection of reality. However, suppose I tell you that two particular children each have an activity level of 3 (very active). You may safely conclude that this is *not* an exactly accurate reflection of reality, because the two children do not have exactly the same activity level.

There are two possible sources of inaccuracy in measurement. First, a numerical value may be inaccurate because insufficient care was used in arriving at the number. Suppose, for example that I have a crate of apples and I say that there are 47 apples in the crate. This number may be inaccurate unless I have carefully counted the apples, but I could get an accurate value by being careful. In the same way, I could "measure" a child's average activity level by watching him for 5 minutes on one day (insufficient care) or by observing him for several days. In the latter case my judgment is likely to be a more accurate reflection of his average activity level. This type of inaccuracy, although all too common in research in the behavioral sciences, can be eliminated in large part.

The second source of inaccuracy is more troublesome. There is a basic difference between apples and behavior. Apples come in whole units, whereas behavior does not. This basic distinction between *discrete entities* and *continuously variable traits* cannot be avoided. Things, which exist in discrete units, can be accurately represented by numbers. Traits, which show continuous variation, can be measured only within some margin of error. Thus virtually all measurements of interest to behavioral scientists contain some amount of error of this type, and the problem is to minimize it. In general, the techniques for handling measurement error are beyond our area of concern in this book; however, the following section presents some useful background for our later topics.

**Error.**   Virtually every number that is used to represent the amount of some trait is in error to some degree, because the traits are continuously distributed in the group of individuals in which we are interested. A continuous trait is one in which the scale of measurement can be divided into infinitely small steps. For example, length is such a trait. No two things, not even rulers, are precisely the same length. When we say that two people are each 60 inches tall we cannot claim that they possess exactly the same amount of the trait height. We can claim that they are both closer to 60 on the scale than they are to 59 or 61, but if we were able to use a ruler that had divisions of sixteenths of an inch we would probably find differences between them. Even if we did not find differences at the level of precision of the sixteenths of an inch, we would at one thousandth, one millionth, or at some level. Whenever a scale is used to measure a continuous trait, the precision of the scale values is restricted by the degree to which we can make them small.

Let us return for a minute to our example of activity level in children. We may reasonably consider activity level to be a continuous trait in which individuals can vary in infinitely small steps from essentially no activity to extreme activity. When we assign the scale values 1, 2, or 3 to particular individuals we are saying that those children who receive a scale value of 1 are more like each other in activity than they are like the children who are assigned a scale value of 2. In fact, what we are usually doing in cases like this is taking the mean of low activity children and assigning that mean a scale value of 1. Then, the mean activity levels of children judged moderately active and extremely active are assigned scale values of 2 and 3, respectively. Thus each child is given the mean (or some other central value) of the group of which he is a member as his scale value.

The mean of the group has certain advantages that make it a good summary value for representing the group. The prime advantage from our point of view is that if we knew each group member's exact value on the trait we could tell how much error was involved in using the mean to represent him, and the average squared error across the whole group would be smaller than for any other possible value. That is, if we subtracted the group mean from a person's true value, this difference would be the error involved in using the mean to represent that person. If we do this for each person, square the resulting errors, and sum them for the whole group, we will find the amount of error to be smaller from using the mean than from using any other central value. Thus the mean of the group gives us a minimum error value to represent all members of the group. This principle applies regardless of whether we define the group as those individuals who obtained a particular scale value or as those falling within some range of scale values. We shall see in Chapter 2 that this principle of using group means to minimize the error involved in measurement will also give us the best results when we wish to predict one variable from another. (The reader should recognize that the mean of these squared deviations from the group mean is the group's variance. The variance, then, is in a sense a measure of the average error made in using the mean to represent each individual in the group.)

## SHAPE OF RELATIONSHIP

The third important aspect of our data which we must consider is the nature of the relationships between variables. We have already discussed the problems of the meaning of our numbers as representatives of real events and the accuracy of this representation. Now we must look at the pattern into which the numbers fall.

This is a book about the relationships between variables. We will generally not be concerned with the distributions of scores for single variables. These will be assumed to be normal except where otherwise stated. Normality is an assumed property of the data in the product

**TABLE 1-1**

*Height and Weight Scores for Ten Hypothetical Individuals.*

| Individual | Height (in.) | Weight (lb) |
|:---:|:---:|:---:|
| 1 | 74 | 200 |
| 2 | 65 | 130 |
| 3 | 70 | 155 |
| 4 | 68 | 170 |
| 5 | 63 | 119 |
| 6 | 71 | 182 |
| 7 | 72 | 210 |
| 8 | 66 | 145 |
| 9 | 60 | 115 |
| 10 | 68 | 140 |

moment correlation and generalizations from it, although it is not certain how important it is that the assumption is met. There is some evidence (Terwilliger, 1970, personal communication) that even extreme deviations from normality do not have much affect on the results of a factor analysis.

**The Scatter Plot.** When we are examining the relationship between two variables it is often very useful to make a picture of the data so that we can see the relationship directly. One way of doing this is to use a pair of coordinate axes to represent the two variables and show each person as a point located by his scores on the two variables.

We can use the height and weight data presented in Table 1-1 as an example. The scatter plot representing these data is given in Figure 1-1, where the vertical axis corresponds to the weight variable and the horizontal axis represents height. The scale of measurement is marked off on each axis as an aid in finding the proper point for each person. Later we will omit the scales and the identification of each person so that we will not clutter up the plot, but for now they will help the reader to grasp the technique of scatter plots.

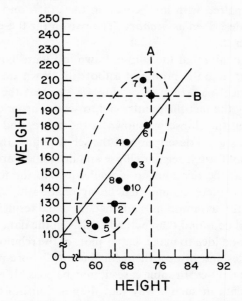

*Figure 1-1. Scatter plot of height and weight data for ten individuals from Table 1-1, weight predicted from height.*

Each person in our group is represented as a point which is the intersection of his values on the two scales. Thus individual 1 has a height of 74 inches and a weight of 200 pounds, and the point representing him is located where lines drawn through his scale value for each variable and parallel to the other axis intersect (lines *a* and *b*). In the same manner, the point for individual 2 is located at the intersection of the lines representing 65 inches and 130 pounds. The points for the other eight people are located in the same manner. Precisely this same approach is used to prepare scatter plots showing the relationship between any two variables.

Looking at the points in Figure 1-1 we can see that there is a certain regularity in the data. The points tend to form an ellipse, with higher values on one variable tending to be associated with higher values on the other variable. In fact, there is a consistent tendency for the scores on one variable to increase as scores on the other variable increase. When the points scatter in this general pattern, that is, high scores on one variable associated with high scores on the other and low scores appearing with low, we say that the relationship is *positive*. On the other hand, when the scatter plot shows a general tendency for high scores on one variable to be paired with low scores on the other, and vice versa, the relationship is described as *negative*. The reason for these terms become clear in Chapter 2.

We may also note that the ellipse drawn in Figure 1-1 is straight and regular in shape, like the profile of a football. When the points tend to fall in an elliptical pattern such as this, we say that the relationship is *linear*, reflecting the fact the points tend to scatter around a straight line running through the ellipse, as shown in the figure, and that a straight line can give us a good description of the set of points. In cases in which the points do not form a regular ellipse and do not fall around a straight line, we say that the relationship is *nonlinear*. All the topics treated in the following chapters assume that linear relationships prevail among all variables. If this assumption is not met, the usual result is that a smaller relationship will be found than actually exists in the data. Consequently, it is usually a good idea to make scatter plots of the relationships between pairs of variables and inspect them for linearity before proceeding with more complex analyses. In some cases it may be possible and permissable to rescale the data in such a way that the relationship becomes linear. Once again, we will have more to say on this topic in later chapters.

Outlines of scatter plots illustrating two types of *nonlinear* relation-ship are presented in Figure 1-2. The first (Figure 1-2*a*) is a nonlinear negative relationship in which there is a general tendency for higher values of the variable called *Y* to be associated with lower values of the variable called *X* in the left part of the plot, and this tendency is reversed in the right part of the plot. Figure 1-2*b* illustrates a nonlinear relation-ship that is generally positive but becomes negative for the highest values of *X*. This general shape of scatter plot (Figure 1-2*b*) is frequently seen in two types of situations: in learning and motivation studies in which very high levels of motivation lead to lower learning performance (other-wise known as the Yerkes–Dodson effect) and in studies of intelligence and performance in repetitive tasks where very bright individuals get bored with the task and do not perform as well as they could.

One unfortunate result of the use of computers in the behavioral sciences is that researchers are less likely to prepare scatter plots of their

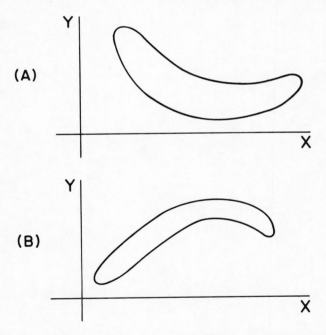

*Figure 1-2   Scatter plots of two types of nonlinear relationships. Part (a) is a generally negative relationship. Part (b) shows a generally positive relationship.*

data than was the case 20 or 30 years ago. There are ways to test the linearity of a relationship without inspecting the scatter plot (which we deal with in the next chapter), but there is no substitute for being close to the data and "getting your hands dirty." In this way the behavioral scientist can get a much better feel for his data. It is even possible to have the computer prepare the scatter plots. Although there are definite restrictions on the plot because of the inflexibility of line printers, these plots are better than no plots at all.

I will close this chapter with a plea. Look at your data. Do not rely on summary indices to tell you what is going on. These indices are useful, but there is no substitute for first-hand familiarity with the data.

# 2

## SUMMARY INDICES
## FOR
## BIVARIATE RELATIONSHIPS

When reporting the results of research, the scatter plot is too inefficient, particularly if there are many variables involved. What is needed is an index that can tell the person reading the report what the scatter plot would look like. Verbal descriptions such as "the scatter plot was long and skinny and went from the lower-left corner to the upper-right corner" are obviously not satisfactory because they are too subject to individual interpretation as to what is meant by such phrases as "long and skinny." What is needed is a numerical index that will accurately describe the type and amount of relationship between two variables. In this chapter we discuss several of the indices that have been used to summarize the relationship between two variables.

Let us consider the hypothetical measure of activity level we described in Chapter 1. We assigned each child an activity level score based on some observations of his behavior. Suppose that we also kept a count of the number of "mishaps" in which each child was involved. This might include objects broken, fights, injuries, and the like. From such a study we might obtain the data in Table 2-1.

Questions concerning the amount of relationship between these two variables can be phrased in several ways. For example, if there is a relationship between activity and mishaps, then knowing the activity level of a child should give us a better idea of the number of mishaps in which he is likely to be involved than we would have if we did not know his activity level. A statement of this kind can be used for two purposes, to predict the probable number of mishaps for a given individual and to describe a group of individuals in terms of the degree to which two variables "go together" in the group. The index called $\eta$ (eta) or the correlation ratio can be used for both of these purposes, and we deal with it here because it is one of the most general approaches to the problem.

The data in Table 2-1 have been arranged by activity level, and arbitrary numbers have been assigned to the individuals. A scatter plot of these data is shown in Figure 2-1. The steady progression of the groups from the lower-left corner to the upper-right corner indicates that individuals with lower activity levels tend to have fewer mishaps than do

**TABLE 2-1**

*Scores on Activity Level and Number*
*of Mishaps for 15 Hypothetical Children.*

| Individual | Activity Level (X) | Number of Mishaps (Y) |
|:---:|:---:|:---:|
| 1 | 1 | 0 |
| 2 | 1 | 1 |
| 3 | 1 | 0 |
| 4 | 1 | 3 |
| 5 | 1 | 2 |
| 6 | 2 | 3 |
| 7 | 2 | 3 |
| 8 | 2 | 4 |
| 9 | 2 | 2 |
| 10 | 2 | 5 |
| 11 | 3 | 4 |
| 12 | 3 | 6 |
| 13 | 3 | 5 |
| 14 | 3 | 6 |
| 15 | 3 | 7 |

individuals with higher activity levels (a positive relationship in terms of what we said in Chapter 1). However, there is some overlap between groups (some individuals in group 1 had more mishaps than some in group 2 did) and there is variability within each group. We would like to know how strong the relationship we observed really is.

The way to attack this problem is to use the mean of each group, as discussed in Chapter 1. If we want to make a general statement about the scale value of a group of observations that will be most nearly true for all members of the group, that is, have the smallest amount of average error for each member of the group, then our best statement will be to give the mean of the group. The mean is the point in a distribution of scores about which the sum of squared deviations are a minimum, and since a deviation of the true value for a person from the stated or predicted value is an inaccuracy or error, using the group's mean as the stated value will lead to the average error being smallest. In Figure 2-1 the mean for each group is given and represented by a short horizontal line. (Notice that we have changed things slightly from Chapter 1. Here group membership is defined by one variable, and our statement is for the other variable. However, the mean number of mishaps for the group is still our best or most accurate general statement about all members of the group. This is true whether we wish to describe the group or predict the most probable number of mishaps for a new member of the group.)

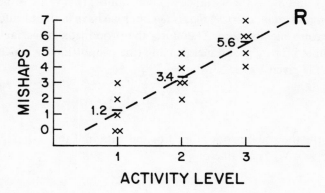

*Figure 2-1   Plot of the relationship between the categorical variable of activity level and the continuous variable of number of mishaps. Note the progression of group means along the straight line "R" and the variability among group means.*

But the question we are really asking is whether knowing which group an individual is in will give us a better estimate of his score on the other variable. In other, more general terms, we are asking whether knowledge of an individual's score on one variable reduces our uncertainty about his score on the other variable. This idea is one important key to understanding all types of correlational procedures. The present example deals with the particular situation in which groups are given, but the logic underlying this example may be generalized to the case in which there are infinitely many possible values for each variable and from there to the case in which we are concerned with many variables simultaneously. The various indices of relationship which we discuss give us information about the degree to which our uncertainty is reduced.

In Figure 2-1 the means for the three groups are 1.2, 3.4, and 5.6, respectively, and the mean of all individuals is 3.4. The variance within each group is the amount of error we would incur if we used each group's mean as our best guess, and the variance of the total sample is the error that would result if we had no knowledge of group membership. The degree to which error remains, even when we know group membership, is given by the ratio of these variances

$$\frac{S^2_{Y(X)}}{S_Y{}^2} = \frac{\text{variance within groups}}{\text{total variance}}$$

where $S^2_{Y(X)}$ is the variance of $Y$ values (mishaps) within a group of $X$ (activity), and $S_Y{}^2$ is the variance of all $Y$ values. This value is the *proportion of total variance (error) that is left* after we have used our information about group membership. Therefore, the proportion of variance which is accounted for by group membership (the proportion by which error is reduced) is given by

$$\eta^2 = 1 - \left(\frac{S^2_{Y(X)}}{S_Y{}^2}\right) \tag{2.1}$$

$\eta$ is simply the square root of $\eta^2$. For the data in Table 2-1, $S^2_{Y(X)} = 1.15$ and $S_Y{}^2 = 4.37$. The value of $\eta^2$ is

$$\eta^2 = 1 - \left(\frac{1.15}{4.37}\right)$$

$$= 1 - .26$$

$$= .74$$

$\eta = .86$. This means that the accuracy of our statements is increased (error reduced) by 74% by knowing to which group an individual belongs.

The statistic $\eta$ by itself is not particularly useful, because it has no direct meaning. It is $\eta^2$ that gives the degree to which knowing $X$ reduces our uncertainty or error in $Y$, or the degree of relationship between the two variables. However, $\eta$ is on the same scale as the product moment correlation coefficient (which we discuss shortly) and is, therefore, of some value for comparing the two indices.

## THE PRINCIPLE OF LEAST SQUARES

The most basic concept on which correlational procedures are founded is the principle of least squares. This is true regardless of whether the objective of a particular analysis is the prediction of scores or the description of a relationship.

Least squares is a direct outgrowth of the concept of variance in a distribution of scores. From the descriptive statistics discussed in Chapter 1 we know that the mean of a distribution is the point in the distribution about which the sum of squared deviations is a minimum and that the variance is the average of the squared distance of all individuals from the mean. If any point other than the mean were chosen, the average of squared deviations from that point would be greater than the variance. Thus if a single value must be taken to describe the entire group of scores, the principle of least squares dictates that the mean is the point to use. It is the point that gives the most accurate prediction *on the average for the group*, because the average squared difference between actual and predicted scores will be smallest, and it is most descriptive of the group as a whole. This is true regardless of the shape of the distribution; however, when the distribution is decidedly nonnormal, other descriptive statistics may be more appropriate for some other purposes.

The principle of least squares can be applied directly to the problem illustrated in Figure 2-1 and Table 2-1. Here we have some additional information we may use to get more accurate prediction or description. If we choose not to use information about activity level in our statements about mishaps, then the best statement we can make for the group as a whole is that each individual will be involved in 3.4 mishaps. If we use this figure as our best guess, the average squared distance between our predicted value and the actual value for each individual is 4.37. That is, on the average the *squared* error of our prediction (or description) is 4.37.

Now, suppose we decide to include information about activity level in making our predictions. Then we have three different subgroups, each with its own mean, and the principle of least squares tells us that we should use the mean for each subgroup as the predicted value for members of that subgroup. The variance of scores within each subgroup about the subgroup's mean is now the average squared error of our prediction, and the average of these subgroup variances is the average squared error of prediction for the group as a whole. This is the value $S^2_{Y(X)}$ in the numerator of the $\eta^2$ formula. As $S^2_{Y(X)}$ becomes small, there is less error in our prediction. Activity level gives us more information about mishaps. The ratio of this residual error (after using our additional information on activity level) to the error without the second variable $(S_Y{}^2)$ is the proportion of variation in scores left unexplained, and $\eta^2$ gives the proportion by which our error was reduced.

This same principle of least squares, that is, minimizing our errors of prediction or description, is used in determining the regression line in bivariate correlation, the regression weights in multiple and canonical correlation, the initial position of factors in some types of factor analysis, and in other correlational procedures. As each of these topics arises we will see how the principle applies.

## GENERAL FORM OF THE CORRELATION RATIO

To recapitulate, we have three different sorts of variation among our $Y$ scores in Table 2-1. There is the total variation of the $Y$ scores around the overall mean of the $Y$s $(S_Y{}^2)$, in this case 4.37 units. There is variation among the $Y$ scores within each group (level of $X$) around the mean of that group $(S^2_{Y(X)})$. It is the average of these "within groups" variances that forms the numerator of the ratio which yields $\eta^2$. The third type of variation is variation among the means of $Y$s for the several groups. It is this variation among the $Y$ means that is said to be variance associated with $X$. The reason for this claimed association is that the $Y$ mean for any group (as defined by its $X$ score) is a constant and is therefore completely predictable. Variation among the $Y$ means $(S_{\bar{Y}}{}^2)$ is thus said to be due to or associated with variations in the values of $X$.

The total variation in $Y$s $(S_Y{}^2)$ is composed of these other two types of variation. In fact, we may write the following equation

$$S_Y^2 = S_{\bar{Y}}^2 + S_{Y(X)}^2 \qquad (2.2)$$

to illustrate the fact that the total variation in $Y$s is the sum of the variance that is associated with $X$ and the variance that is not associated with $X$.

If we represent the difference between our best guess about the individual's score and his actual score by an expression of the general form $(Y_i - Y_i')$ ($Y_i$ is the person's actual score and $Y_i'$ is his predicted score), we see that for the situation in which we have no knowledge about the person's group membership, we get $(Y_i - \bar{Y})$, where $\bar{Y}$ is the overall mean, as an expression of the difference between the person's predicted score and his actual score. (Remember that we predict the overall mean for everybody when we have no information about group membership.) This value represents the error in our prediction. By taking the value of this expression for each person, squaring, and summing over individuals, we get

$$\sum_{i=1}^{N} (Y_i - \bar{Y})^2$$

When this sum of squared deviations from the mean is divided by the sample size, we have

$$\frac{\sum_{i=1}^{N} (Y_i - \bar{Y})^2}{N}$$

which is the expression for the variance of $Y$ scores, $S_Y^2$.

Suppose we now include information about group membership in our attempt to predict a person's score. To do this, we would find the mean of $Y$ scores for the members of each separate group on $X$. Let us say that there are $J$ ($=3$ in our example) groups formed on the $X$ variable. Thus each group has a mean, $\bar{Y}_j$, which is the mean $Y$ score of those people in group $j$ of $X$. Then to make use of the $X$ information in making a guess about a person's score on $Y$, we use $\bar{Y}_j$ for members of group $j$ because this is the value which will have the smallest sum of squared discrepancies between actual and predicted values of $Y$ for the members of that group. For each group on $X$ we have the sum of squared errors as

$$\sum_{i=1}^{N_j} (Y_{ji} - \bar{Y}_j)^2$$

where the summation to $N_j$ is used to show that the number of observations in the groups need not be the same. If we sum these several sums across groups, we have

$$\sum_{j=1}^{J} \left[ \sum_{i=1}^{N_j} (Y_{ji} - \overline{Y}_j)^2 \right]$$

as the total amount of squared deviation from prediction. When we divide this quantity by the total number of people observed the result is a variance, but this time it is the variance of $Y$ scores given the $X$ information, $S_{Y(X)}^2$.

The reader may be a little worried that $S_{Y(X)}^2$ is in some way not a variance. Suppose we consider just one group at a time. We can find the variance of $Y$s in a particular group by the expression

$$S_{Y(X_1)}^2 = \frac{\sum_{i=1}^{N_1} (Y_{1i} - \overline{Y}_1)^2}{N_1}$$

where the subscript 1 is used to denote that these computations use only information from group 1 of $X$. This can be seen to be an ordinary variance of the same form as $S_Y^2$. But, we need to know the average amount of error in prediction for all subjects combined. We can do this by taking a weighted average (weighted by the number of individuals in each group) of the $S_{Y(X)}^2$s for the several groups, or, we can get the value we need directly by the formula

$$S_{Y(X)}^2 = \frac{\sum_{j=1}^{J} \left[ \sum_{i=1}^{N_j} (Y_{ji} - \overline{Y}_j)^2 \right]}{\sum N_j} \tag{2.3}$$

in which $\sum N_j$ is the total number of subjects. The result of these computations, done either way, is the average of the squared discrepancies between observed and predicted scores for all individuals in the study.

We can view a person's observed score $Y_{ji}$ (for person $i$ of the $j$th group) as being made up of three parts: a part due to the total mean, a part due to his group's mean as it deviates from the total mean, and his own deviation from his group's mean. These effects are shown graphically in Figure 2-2, in which the distance labeled 1 is the total group effect, the distance labeled 2 is the group mean effect, and 3 refers to the person's deviation from his group's mean. In equation form, this situation may be expressed as

$$Y_{ij} = \overline{Y} + (\overline{Y}_j - \overline{Y}) + (Y_{ij} - \overline{Y}_j)$$

That is, a person's raw score, $Y_{ji}$, is composed of the total group's level $(\overline{Y})$, plus the deviation of his group from the mean of the total group, $(\overline{Y}_j - \overline{Y})$, plus his deviation from his group's mean $(Y_{ji} - \overline{Y}_j)$. In the same fashion, a person's deviation from the total mean may be expressed by subtracting $\overline{Y}$ from both sides of the equation.

$$(Y_{ji} - \overline{Y}) = (\overline{Y}_j - \overline{Y}) + (Y_{ji} - \overline{Y}_j) \qquad (2.4)$$

The left-hand term is the one we have seen associated with $S_Y{}^2$, and the second term on the right we recognize as the fundamental quantity from

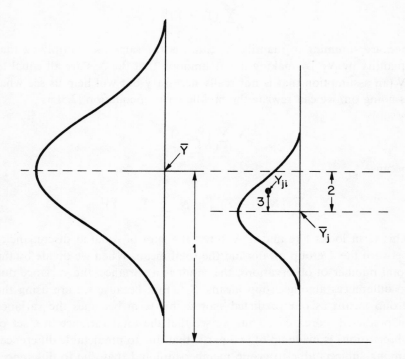

*Figure 2-2   Components of individual scores. Distance 1 is the total group mean* $(\overline{Y})$, *distance 2 is the deviation of the subgroup mean* $(\overline{Y}_j)$ *from the total group mean, and distance 3 is the deviation of the individual's score* $(Y_{ij})$ *from the subgroup mean.*

which we get $S^2_{Y(X)}$. The first term on the right is new, but the reader probably has guessed that it has something to do with the $S_{\bar{Y}}^2$ term from equation (2.2), and this is in fact the case. If we square each of the terms in (2.4) and sum across all subjects in all groups we get

$$\sum_{j=1}^{J} \sum_{i=1}^{N_j} (Y_{ji} - \bar{Y})^2 = \sum_{j=1}^{J} \left[ \sum_{i=1}^{N_j} (\bar{Y}_j - \bar{Y})^2 \right] + \sum_{j=1}^{J} \left[ \sum_{i=1}^{N_j} (Y_{ji} - \bar{Y}_j)^2 \right] \quad (2.5)$$

in which the first and last terms are familiar. In the second term,

$$\sum_{j=1}^{J} \left[ \sum_{i=1}^{N_j} (\bar{Y}_j - \bar{Y})^2 \right]$$

we see that the quantity $(\bar{Y}_j - \bar{Y})^2$ will be a constant for all $N_j$ individuals in group $j$. Thus, we may write $N_j(\bar{Y}_j - \bar{Y})^2$ for

$$\sum_{i=1}^{N_j} (\bar{Y}_j - \bar{Y})^2$$

because summing a quantity $N_j$ times is the same as multiplying that quantity by $N_j$. By making the assumption that the $N_j$s are all equal to $N$ (an assumption that is not really necessary but will help us see what is going on) we can rewrite the middle term of equation (2.5) as

$$N \sum_{j=1}^{J} (\bar{Y}_j - \bar{Y})^2$$

because

$$\sum_{j=1}^{J} [N(\bar{Y}_j - \bar{Y})^2] = N \sum_{j=1}^{J} (\bar{Y}_j - \bar{Y})^2$$

This term looks like and is $N$ times the sum of squared discrepancies between the $J$ group means and the total mean. When we divide by the total number of observations, the result is a variance, the variance due to differences among group means, $S_{\bar{Y}}^2$. But, because we are using the group means as our predicted scores, this term becomes the variance of predicted scores, $S_{Y'}^2$. Thus we see that the total variance in a set of observations is made up of two parts, that due to predictable differences among subjects (due to group membership) and that due to differences that cannot be predicted. The square root of the unpredictable portion of the variance in a set of observations has been given the name standard error of estimate, about which we have more to say later.

Earlier we said that $\eta^2$ is the ratio of predictable variance to total variance (that proportion of the total variance which is predictable), and we defined $\eta^2$ as

$$\eta^2 = \frac{S_{Y'}^2}{S_Y{}^2} = 1 - \frac{S_{Y(X)}^2}{S_Y{}^2}$$

By substitution of the above terms in this equation we get

$$\eta^2 = \frac{\dfrac{\sum\limits_{j=1}^{J} [N_j(\overline{Y}_j - \overline{Y})^2]}{\sum N_j}}{\dfrac{\sum\limits_{j=1}^{J}\left[\sum\limits_{i=1}^{N_j} (Y_{ji} - \overline{Y})^2\right]}{\sum N_j}} = 1 - \frac{\dfrac{\sum\limits_{j=1}^{J}\left[\sum\limits_{i=1}^{N_j} (Y_{ji} - \overline{Y}_j)^2\right]}{\sum N_j}}{\dfrac{\sum[\sum(Y_{ji} - \overline{Y})^2]}{\sum N_j}}$$

which reduces to

$$\eta^2 = \frac{\sum\limits_{j=1}^{J} [N_j(\overline{Y}_j - \overline{Y})^2]}{\sum\limits_{j=1}^{J}\sum\limits_{i=1}^{N_j} (Y_{ji} - \overline{Y})^2} = 1 - \frac{\sum\limits_{j=1}^{J}\left[\sum\limits_{i=1}^{N_j} (Y_{ji} - \overline{Y}_j)^2\right]}{\sum\limits_{j=1}^{J}\sum\limits_{i=1}^{N_j} (Y_{ji} - \overline{Y})^2} \tag{2.6}$$

In analysis of variance terminology this is the ratio of the sum of squares between groups to the total sum of squares, or 1 minus the sum of squares within groups divided by the total sum of squares. Thus, again, $\eta^2$ is the proportion of the total variance that is due to differences between group means or the proportion of variance that is predictable from knowing to which group the individuals belong.

The use of $\eta$ requires that we make some assumptions about the properties of our measurement. We assume that the variable to be predicted $(Y)$ is measured on an interval scale $(X$ can be a nominally measured variable). This is necessary so that we can justify the computation of means and variances. We also make the assumption that the variances of the $Y$s in the several groups formed by the $X$ variable are equal (i.e., $S_{Y(X_1)}^2 = S_{Y(X_3)}^2$, etc.). This assumption is necessary so that $S_{Y(X)}^2$ for the whole study accurately reflects the variance within each group. If we wish to test the statistical significance of $\eta^2$, we must make the additional assumptions that our groups represent random samples from populations of individuals who would be classified into the several groups on $X$ and that the distribution of $Y$ scores within each group is normal.

The test for the statistical significance of $\eta^2$ is given by the formula

$$F = \frac{\eta^2/(J-1)}{(1-\eta^2)/(N-J)}$$

where $J$ is the number of groups formed from the $X$ variable and $N$ is the total number of individuals in the study. This $F$ has $J-1$ numerator degrees of freedom $(df)$ and $N-J$ denominator $df$. The value of $F$ is compared to the appropriate value in an $F$ table, and $\eta^2$ is statistically significant if the value of $F$ obtained from the data is greater than the tabled value. Finding $\eta^2$ to be statistically significant means that there is greater than chance improvement in our ability to predict $Y$ scores using knowledge of $X$ over our ability to predict without $X$. Such a finding also means that the proportion of variance in $Y$ scores due to group membership on $X$ (i.e., the variance due to differences among the $\overline{Y}_j$s) is greater than zero. We conclude that there is a real, nonzero relationship between $X$ and $Y$.

Those who have had some introduction to correlation will have noticed that we have not assumed any particular form for the relationship between $X$ and $Y$. The correlation ratio makes no assumptions about the form of the relationship. In fact, we made no assumptions at all about the measurement properties of the $X$ variable. This means that $\eta^2$ will give us a measure of association for any shape of relationship, linear or curvilinear. All that matters is that we be able to compute means and variances of $Y$ for the several groups of $X$. The correlation ratio $(\eta^2)$ will give us a measure of the strength of relationship that is independent of the form of that relationship.

### CONTINUOUS, INTERVAL X

The situation we have just described is given in terms of a nominally measured $X$ variable. It is often the case that we have two variables that are measured on an interval (or ratio) scale, for example, the data presented in Table 1-1 and Figure 1-1. When data of this type are represented in a scatter plot, the results may form a regular ellipse, such as the points plotted in Figure 1-1, or they may form a twisted elliptical figure like one of those in Figure 1-2. If we do not place a restriction of linearity on the form of the relationship, we must be prepared to handle results of the latter type. We accomplish this by a slight modification in our use of $\eta^2$.

When our observations have been recorded on an interval scale of measurement we can, if we wish, form groups by breaking up our interval scale into ordered categories, which is exactly what we do in forming a frequency distribution. (Remember that each score value is really a category on the continuous scale.) Given a continuously measured $X$ variable, we can form arbitrary groups on the variable and compute an $\eta^2$ in the same manner as though the groups were discrete. The value that results from such an analysis is an index of relationship between the two variables which requires no assumptions about the form of the relationship. Of course, the categories of $X$ are ordered so that we cannot arrange them at will, as would be possible with a nominally scaled $X$, but this does not disturb the correlation ratio. However, some information contained in the continuous $X$ variable is discarded in the process of forming groups for the $\eta$ analysis. Our knowledge of a person's position on the scale is not as precisely specified by his group membership as it is by his score. Of course, because the scores are really categories, we could use them for an $\eta$ analysis; however, in most cases there are not enough individuals at each score value to give good estimates of the within groups variances. Also, the analysis becomes more complex when we have a very large number of groups; thus it is common practice to form groups containing several score values of the predictor variable for performing an $\eta$-analysis. We lose some information to gain simplicity.

This problem of loss of information may be partially overcome in two ways in actual situations. First, there is the technique of polynomial curve fitting, in which the object is to find a line definable by some polynomial equation that fits the scatter plot of the data in a least squares sense. This approach is not used very frequently in the behavioral sciences; thus we will not discuss it except to note that the topic is treated by Ezekiel and Fox (1959) and Draper and Smith (1966). The second approach to using the maximum amount of information is to place the restriction of linearity on the form of the relationship. This approach, in fact, throws away nonlinear information, but when the assumption of linearity is not too seriously violated, the gain in descriptive power is well worth the price.

The assumption of linearity of relationship means that the group means of data prepared for $\eta$ analysis would all fall on a straight line and that a straight line gives the best least-squares fit to the data in the scatter plot. We will have a great deal more to say about linear relationships. In fact, most of the remainder of this book deals with various elaborations of the basic idea of linear least-squares fit. However, before we plunge into the myriad beauties of the linear relationship, let us take a look at a hypothetical situation that may help us get ourselves ready for the task.

Suppose that we have a continuous $X$ and $Y$ and that we have a very large number of people from whom we have gathered our $X$ and $Y$. Since $X$ is continuous, it is theoretically possible to have the units of measure on $X$ get infinitely small. However, there are practical limits to the size of the units on $X$ which are conditioned solely by our inability to make distinctions beyond some given point. Thus, although our $X$ is really continuous, we make it into a discrete series at some point in our measurement. The same applies to $Y$. (The reader who still doubts this contention might consider the variable of time. At what point does the scale for the measurement of time become continuous, at the second, the millisecond, the microsecond, the nanosecond? Obviously, each of these measuring units forces a continuous variable into a discrete series of units whose size is limited by our measuring device.)

Now that we have all these people and their data, we can proceed to determine the nature of the relationship between $X$ and $Y$. First, let us break $X$ up into some rather large categories in preparation for computing $\eta^2$. The results might look something like those in Figure 2-3$a$, in which the $Y$ means for the distributions within each category of $X$ are connected by the solid line.

Let us consider what would happen if we started making our categories of $X$ smaller and smaller and more and more numerous. By doubling the number of categories we would get a situation like that shown in Figure 2-3$b$. Parts $c$ and $d$ of Figure 2-3 show the situation when the number of categories is 12 and 16, respectively. The reader should notice that the angles in the line connecting the $Y$ means of the several groups on $X$ become less severe as the $X$ variable is broken into smaller and smaller categories. Remembering that this is an ideal situation (i.e., one in which truth and beauty prevail and one in which the data always fulfill the statistician's assumptions), we can continue making finer and finer distinctions on $X$ while still being able to compute means of the $Y$ scores of the several individuals who are in each category. As

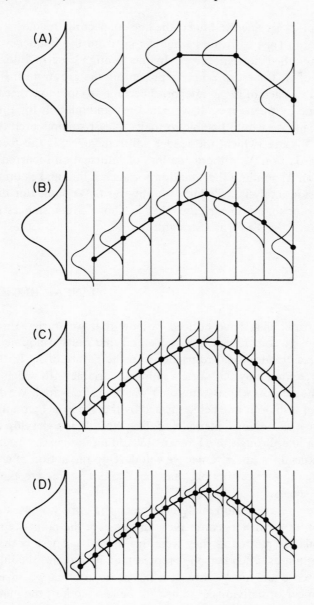

*Figure 2-3  Effect of having finer gradations of X. The curve in the line connecting the means becomes smoother as the number of categories of X increases.*

the categories get finer and finer, the line connecting the means becomes more smooth (just as it does when we go from the frequency polygon to the smoothed normal curve). In the limiting case of infinitely fine divisions of $X$, the set of points representing the group means of the $Y$ arrays within each of the $X$ categories become so close together that they form a line by themselves. If we could get an equation for this line of group means we would be able to make a least squares prediction of a person's $Y$ score (which, the reader will remember, is the mean of $Y$s for that group on $X$) without the loss of information incurred by categorizing on $X$. Finding the equation for such a line makes up the topic of regression analysis, to which we now turn. We limit our discussion of regression to the linear case, that is, those circumstances in which the means of the $Y$s form a straight line.

## LINEAR REGRESSION

The reader may have noticed by now that we started this chapter with a loosely constrained $X$ variable and that our development has involved placing additional restrictions on the nature of $X$. In the beginning we required only that $X$ be a nominal variable with which we could classify the subjects whose status on $Y$ was being measured. We developed the idea of a smooth line being formed by the $Y$ means by requiring that $X$ become a continuous interval variable. We will now develop the equation for a line through the $Y$ means by adding the next restriction, that the relationship is linear. Later we will develop the notion of correlation by adding another restriction, that the $X$ and $Y$ scores are expressed as standard scores.

We begin a regression analysis in much the same way we started our discussion of $\eta$, with a scatter plot. In this case the points representing individuals are plotted as they were in Figure 1-1 where both $X$ and $Y$ were continuous. Each person is represented by a point, the intersection of his $X$ score and his $Y$ score. It is not necessary that we have the very large number of individuals we had in the last section; the mathematics of regression analysis will let us find the best fit regression line without finding group means. However, if our assumption of the linearity of the relationship is precisely met, the line will go through the points where the group means would have been if we had computed them. We use the regression line as a substitute for the group means.

## DEFINING THE LINE

To get an equation that uniquely defines a straight line we need to know two things about the line. First, we need to know how quickly the line rises or falls. This is called the *slope* of the line. Second, we need to know the point where the line crosses the vertical ($Y$) axis of our scatter plot. The numerical value of this point is called the *intercept* of the line. The slope of the line tells us the direction of relationship in the scatter

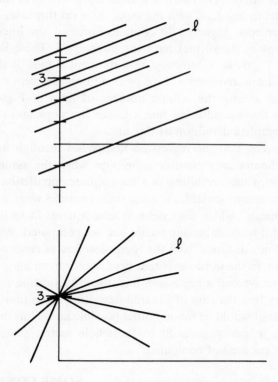

*Figure 2-4   The slope and intercept define the regression line. The upper part of the figure shows lines with the same slope and different intercepts, while the lower part shows lines with the same intercept and different slopes.*

plot, because it is defined as the number of units of increase in $Y$ for each unit of increase in $X$. When low scores on $X$ are associated with low scores on $Y$ and high scores on $X$ are associated with high scores on $Y$, there is a *positive increase* in $Y$ with increasing $X$ and the value of the slope is positive. In the case of low scores on $X$ being associated with high scores on $Y$, and vice versa, there is a *negative increase* (a decrease) in $Y$ with increasing $X$ and the sign of the slope is negative. This is why we speak of positive and negative relationships. The intercept tells us the value of $Y$ when $X$ is zero.

We can see that both these pieces of information are necessary and that the two of them are sufficient to define the line by looking at Figure 2-4. In Figure 2-4$a$ there are several lines, all with the same slope but with different intercepts. There are infinitely many lines that could be drawn parallel to line $l$, all with the same slope (in this case, .5) but with different intercepts. Figure 2-4$b$ contains some of the lines (there are infinitely many of them) that have the intercept 3. These lines all pass through the $Y$ axis at 3, but they have different slopes. If the equation of our best-fit line required a slope of .5 and an intercept of 3, there is only one line among the infinite number of lines in Figure 2-4 that satisfies both these conditions, line $l$. Given the slope and the intercept we have a complete definition of this line.

We have seen that the regression line passes through the means of the several theoretically possible subgroups when the assumption of a linear relation is met, resulting in a least-squares prediction or description. For continuous variables the line itself provides what we might call a "running mean" which also yields a least-squares fit to the points in the scatter plot when subgroup means are not computed. We may think about a person's distance from the regression line as error of prediction or description, in the same sense that his distance from his group's mean is error. When we find a regression line, then, its location is determined in such a way that the sum of squared deviations from that line (errors) is smaller than it would be for any other possible line. Thus the regression line provides a least-squares fit to the whole scatter plot, even when subgroup means are not computed.

## SOME CONVENTIONS

We have implicitly adopted some rather arbitrary conventions about notation and terminology in our discussion thus far. At this point it is proper to make these conventions and a few others explicit for the reader. First, we have adopted the convention of calling the variable to be predicted $Y$. In a practical situation we may wish to predict either variable from the other. The label of $Y$ for the predicted variable is purely arbitrary. For example, we may wish to predict achievement from aptitude, or we may wish to predict aptitude from achievement. If we arbitrarily assign the label $Y$ to aptitude and call achievement $X$, then in the former case we would be predicting $Y$ from $X$, whereas in the latter we would be predicting $X$ from $Y$. The equations of the lines for making these two predictions would probably be different. However, for our purposes it is just as easy to shift the labels attached to the variables as to provide two different equations, one for predicting $Y$ from $X$ and the other for predicting $X$ from $Y$. As an example of what will happen when the labels are shifted, the reader should compare the scatter plot in Figure 2-5 with that in Figure 1-1. In the former case height is the predicted variable and weight is the predictor. In the latter (Figure 1-1) the predicted variable is weight and height is the predictor. Notice that the slopes of the regression lines will be quite different for these two cases.

We have also adopted the convention of showing a scatter plot as an ellipse in cases in which we have not plotted a specific set of data. The ellipse is an idealized representation of what the *outline* of the scatter plot would look like. If we actually had such an ideal set of data, the scatter plot would have an outline of the elliptical form, and the points

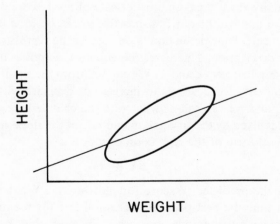

*Figure 2-5   Theoretical scatter plot for predicting height from weight. Note the difference from Figure 1-1.*

representing people would cluster most heavily in the center of the ellipse, spreading out more thinly toward the edges.

There are two generally accepted symbols for the slope and intercept of the regression line that are used from now on. The symbol for the slope is $B$ and that for the intercept is $A$. Many statistics books use $B_{Y \cdot X}$ or $B_{YX}$ to indicate that $Y$ is being predicted from $X$ and $B_{X \cdot Y}$ or $B_{XY}$ to refer to the reverse case. Our convention of calling the predicted variable $Y$ in all cases makes this distinction unnecessary; thus the symbol $B$, when used, will stand for $B_{Y \cdot X}$.

## THE REGRESSION EQUATION

When we use a straight line to make predictions from a continuous variable $X$ to another continuous variable $Y$, we theoretically have a known $X$ and an unknown $Y$. Of course, to develop the equation in the first place we need to know both $X$ and $Y$ for a group of people. Given this information, we can develop an equation that can be used for two purposes. If we have a new individual about whom we have only information on $X$, we can use the regression equation to make a best guess as to his standing on $Y$. Assuming that our new individual is from the same population from which we drew our original sample (an assumption we virtually always make), our estimate of his $Y$ score derived from the regression line will be our best possible guess, because it will tell us his group's mean. Thus prediction is one use of the regression equation. The other is description. The regression equation describes the nature of the relationship between $X$ and $Y$. We can tell from the equation whether there is a tendency for $Y$ scores to increase or decrease with increasing values of $X$, and we can tell at what rate the change takes place. This information is given by the sign and magnitude of the slope coefficient, $B$.

The general form of the regression equation is

$$Y_i' = BX_i + A$$

where $Y_i'$ is the predicted $Y$ score for individual $i$. What we must do now is determine the best possible regression line for the data we have at hand. We wish to minimize our errors in predicting $Y$ scores from $X$ scores using the regression line. This means that on the average across the group, the quantity $(Y_i - Y_i')$ should be a minimum since this is the difference between actual and predicted scores.

At this time the reader should recall two things. First, we have developed the idea of the regression line as the line passing through the means of the infinitely many groups. Second, the sum of deviations from a mean is zero. Therefore, across the group as a whole, the sum of the several errors of prediction will equal zero just as the quantity $\Sigma(X_i - \overline{X})$ equals zero. We could reach this conclusion by a more mathematical route, but the reader who wishes such assurances should consult McNemar (1969) for details of the proof. For our purposes it is sufficient to say that the solution for the regression line involves a least-squares solution for deviations from the regression line.

The derivation of the equation for computing the regression coefficient $B$, although not very difficult to follow, requires a touch of calculus. We give only the final computing formula, which is

$$B_{Y \cdot X} = \frac{N \sum_{i=1}^{N} X_i Y_i - \Sigma X_i \Sigma Y_i}{N \Sigma X_i^2 - (\Sigma X_i)^2} \tag{2.7}$$

and the definitional formula, which involves the deviations of the $Y$s from the $Y$ mean and the $X$s from the $X$ mean,

$$B_{Y \cdot X} = \frac{\sum_{i=1}^{N} [(Y_i - \overline{Y})(X_i - \overline{X})]}{\Sigma(X_i - \overline{X})^2} = \frac{\Sigma y_i x_i}{\Sigma x_i^2} \tag{2.8}$$

where $y_i = (Y_i - \overline{Y})$ and $x_i = (X_i - \overline{X})$. The computational formula is, of course, much easier to use in practice, but the definitional formula shows us that the $B$ coefficient is simply the sum of the crossproducts of the deviation scores divided by the sum of squared deviations of the $X$s. (Starting with equation 2.7, subscripts and ranges of summation are omitted unless needed for clarity.)

We can get an understanding of what the $B$ coefficient tells us by some of the properties of deviation scores. First, deviation scores are signed numbers, some of them positive and some of them negative. Positive deviations are those that fall above the group mean, and negative deviations are those below the mean. The product of the two deviation scores for an individual will be positive if both deviations have the same sign and negative if the signs are not the same. When these crossproducts are summed for all individuals, the signs must be kept in the addition. Thus if most of the pairs of deviation scores are of the same sign, the sum of the crossproducts will be positive and the sign of the resulting B will be positive. When this situation occurs we say that there is a positive relationship between $X$ and $Y$, which means high scores on $Y$ tend to be found when the individuals have high scores on $X$ and that low scores on $Y$ accompany low scores on $X$. A negative value for $B$ indicates that high scores on $X$ have low scores on $Y$, and vice versa. The reader should recognize the quantity $\Sigma x_i^2 = \Sigma(X_i - \overline{X})^2$ in the denominator of (2.8) as being very like a squared standard deviation or a variance. In fact, $\Sigma x^2 = NS_x^2$. Obviously, this quantity is always positive and acts merely as a scaling factor. Its effect is simply to scale $B$ so that it reflects the number of units of gain or loss in $Y$ per unit of increase or decrease in $X$. A value of $B$ of $+.5$ means that the value of $Y$ increases at the rate of 1/2 unit of $Y$ per unit of increase in $X$, whereas a value of $B$ of $-1.6$ tells us that the value of $Y$ decreases by 1 and 6/10 units of $Y$ per single unit of increase in $X$. The nature of the underlying measurement scales must be kept in mind when interpreting the sign of $B$. This interpretation holds only when larger numbers indicate increasing amounts of both of the traits being measured or when smaller numbers reflect larger amounts of both traits. The interpretation must be reversed when large numbers indicate more of one trait and smaller numbers more of the other, as would be the case if we were studying the relationship of age and number of errors in a learning task.

The value of the intercept, $A$, simply fixes the point at which the regression line crosses the $Y$ axis and reflects how far above or below the $X$ axis the scatter of points lies. It is computed by the formula

$$A = \overline{Y} - B\overline{X} \qquad (2.9)$$

showing that it is an adjustment to reflect the relative magnitudes of the two means. It has no necessary function in describing the nature of the relationship between the two variables, but it is necessary for predicting *raw scores* on $Y$ from *raw scores* on $X$.

We will use the data from Table 1-1 to illustrate the computation of the regression coefficient and the intercept and to show how these may be used in computing predicted scores. The computations are given in detail in Table 2-2. From the computations in this table we see that for each inch of increase in height in our sample there is an increase of

**TABLE 2-2**

*Computation of the Regression Equation for Predicting Weight from Height for Data from Table 1-1.*

| Individual | Height (X) | Weight (Y) | XY | Height² | Weight² |
|---|---|---|---|---|---|
| 1 | 74 | 200 | 14,800 | 5,476 | 40,000 |
| 2 | 65 | 130 | 8,450 | 4,225 | 16,900 |
| 3 | 70 | 155 | 10,850 | 4,900 | 24,025 |
| 4 | 68 | 170 | 11,560 | 4,624 | 28,900 |
| 5 | 63 | 119 | 7,497 | 3,969 | 14,161 |
| 6 | 71 | 182 | 12,922 | 5,041 | 33,124 |
| 7 | 72 | 210 | 15,120 | 5,184 | 44,100 |
| 8 | 66 | 144 | 9,504 | 4,356 | 20,736 |
| 9 | 60 | 115 | 6,900 | 3,600 | 13,225 |
| 10 | 68 | 140 | 9,520 | 4,624 | 19,600 |
| | $\Sigma X = \overline{677}$ | $\Sigma Y = \overline{1,565}$ | $\Sigma XY = \overline{107,123}$ | $\Sigma X^2 = \overline{45,999}$ | $\Sigma Y^2 = \overline{256,471}$ |

$$B_{Y \cdot X} = \frac{N \Sigma XY - \Sigma X \Sigma Y}{N \Sigma X^2 - (\Sigma X)^2} = \frac{10(107,123) - 677(1,565)}{10(45,999) - (677)^2} = \frac{1,071,230 - 1,059,505}{459,990 - 458,329} = \frac{11,725}{1,661} = +7.059$$

$$A = \overline{Y} - B \overline{X} = 156.5 - (+7.059)(67.7) = 156.50 - 477.89 = -321.39$$

7.059 pounds in weight and that the regression line crosses the $Y$ axis at $-321.39$. Although it is meaningless to say so, we might also conclude that an individual who is zero inches tall would weigh $-321.39$ pounds. Interpretation of the intercept must be restricted to those situations in which a value of zero on the predictor variable has some meaning, which it obviously does not in this case. (In fact, the relationship between height and weight is nonlinear in the lower ranges of these variables.)

To show that the magnitude of the regression coefficient and the value of the intercept may change when the labels for predictor and predicted variables are changed, we compute the value of $B$ for predicting height from weight. In this case height becomes $Y$ and weight is $X$. The values of weight squared were included in Table 2-2, although we did not use them in our previous calculations. We will use them now. The values of $B$ and $A$ for predicting height from weight are

$$B = \frac{10(107123) - 677(1565)}{10(256471) - (1565)^2} = \frac{+11725}{2564710 - 2449225} = \frac{+11725}{115485} = .1015$$

$$A = 67.7 - (.1015)(156.5) = 67.7 - 15.89 = 51.81$$

The two scatter plots in Figure 2-6 will help give an understanding of why these two regression equations are so different. The first part of the figure shows the scatter plot when weight is being predicted from height. Here the ellipse is almost vertical, indicating a rapid change in weight per unit change in height. In the second part of the figure height has been put on the $Y$ axis as the predicted variable (thus reversing the labels of the variables as discussed above). Now the ellipse is almost horizontal, indicating a very gradual change in height per unit change in weight.

Although the two regression equations and scatter plots are quite different, they can be used in the same way and will give the same relative accuracy. This fact can be most readily seen if we take an individual who lies on the regression line and use his height to predict his weight, and vice versa. Suppose that we have a person who is 69 inches tall and weighs 165 pounds. If we wish to predict his weight from his height, we must use the values for $B$ and $A$ which we computed in Table 2-2. The regression equation gives us

$$Y' = 7.059(69) - 321.39 = 487.06 - 321.39 = 165.67$$

By using the second regression equation, the one which we developed for predicting height from weight, we get

$$Y' = .1015(165.67) + 51.81 = 16.85 + 51.81 = 68.7$$

These values do not agree perfectly, because the individual does not lie precisely on the regression line and because there is rounding error in computing the coefficients. As seen later in our discussion of the standard error of estimate, the *relative magnitudes* of errors in prediction (i.e., the differences between $Y$ and $Y'$ scores) are the same for a given set of data regardless of which of the two regression equations we use.

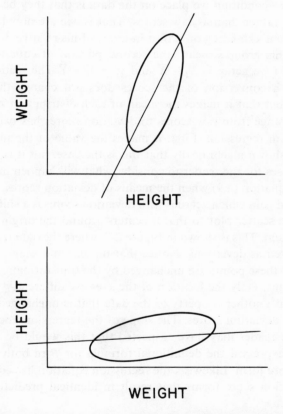

*Figure 2-6    Theoretical scatter plots for predicting weight from height and height from weight. Note the difference in regression slopes and intercepts.*

## REGRESSION COEFFICIENTS WITH DEVIATION SCORES

Thus far in our discussion we have placed two constraints on our predictor variable, $X$, by which we have been able to move from the correlation ratio, which gave us an index of the strength of relationship, to the raw score regression equation, which has allowed us to describe the form of the relation between $X$ and $Y$. These two constraints are an $X$ measured on an interval scale and the requirement that the relationship be linear. By retaining these two conditions and adding one more, we are able to move a little closer to the concept of the correlation coefficient.

The new condition we place on the data is that they be expressed as deviations rather than as raw scores. That is, we are now talking about an individual's standing on a trait in terms of his relative distance above or below his group's mean (the second portion of equation 2.8). For individual $i$ we get $x_i = X_i - \overline{X}$ and $y_i = Y_i - \overline{Y}$. The reader will recall that such a conversion of the scores does not change their standard deviation but that it makes the mean of each distribution zero.

The change from raw scores to deviation scores has two useful consequences in regression. First, it makes the value of the intercept zero. It can be shown algebraically that this is the case, but it is sufficient for our purposes for the reader to consider what will happen in the formula for $A$ in equation (2.9) when the means of deviation scores, 0s, are used. The second gain which accrues with deviation scores is a shift in the location of the scatter plot so that it centers around the origin of the coordinate system. This is shown in Figure 2-7, where the data from Table 1-1 are replotted as deviations. Notice that the size and shape of the ellipse formed by these points are unchanged by the transformation of the data to deviations. Only the location of the axes are different.

There is another property of the data that is unchanged by the conversion to deviation scores. The slope of the regression line remains the same. The reader may have suspected that this would be the case, because we expressed the definitional formula for $B$ in both raw and deviation score form. Likewise, the regression equation has equivalent raw and deviation score forms that result in identical predictions. We use the formula

$$y' = Bx$$

to predict a deviation score, $y'$, from a deviation $x$. Using the data from our previous example, our 69-inch subject has a deviation height of +1.3 inches (69 − 67.7). Therefore, his predicted weight deviation is +9.17 pounds (1.3 × 7.059), which, when converted back to a raw score, is (156.5 + 9.17) = 165.67, the same value that we obtained from our raw score equation.

Notice that to use the deviation score formula we must subtract $\bar{X}$ from $X_i$, put the resulting $x_i$ into the equation to solve for $y_i'$, and then add $\bar{Y}$ to $y_i'$ to get $Y_i'$. Because data seldom become available directly as deviations and because there is no increase in usefulness in the resulting regression coefficient, the deviation score form of the regression equation is seldom used. It does, however, provide us with a useful insight about the regression line. When the regression line is found according to the principle of least squares it will pass through a point whose coordinates are the means of the two variables. This may be seen from the fact that the regression line passes through the origin when the scatter plot is of deviation scores, and this characteristic holds for raw score scatter plots as well. The reader will find this fact not unreasonable when he recalls that the mean of a distribution of scores is also the point in the distribution about which the sum of squared deviations is a minimum. (An algebraic proof is also possible and is given in some introductory statistics books.)

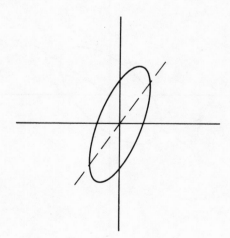

*Figure 2-7   Scatter plot of deviation scores. Note that the origin is in the center of the ellipse and that the regression line passes through the origin.*

## PRODUCT MOMENT CORRELATION

There is one final restriction we may impose on the data which buys us a great deal in terms of what the resulting index permits us to say about the data. This final restriction is that the data be expressed in standard score form. When this condition is met one very important thing happens. The properties of deviation scores are retained, but the process of standardizing (converting raw scores to $Z$ scores) equalizes the variances of the two variables. This has some rather far-reaching consequences, one of which is that the shape of the scatter plot is changed. Comparison of Figure 2-8, in which the data from Table 1-1 are plotted as standard scores, with Figure 2-7, the plot of deviation scores, shows that the ellipse of points has changed its orientation to the coordinate axes and has become shorter. Both of these changes are due to the equalizing of the variances of the two variables. There are several other less obvious changes that take place as a result of standardization, but we will hold off consideration of these until the coefficient itself has been discussed.

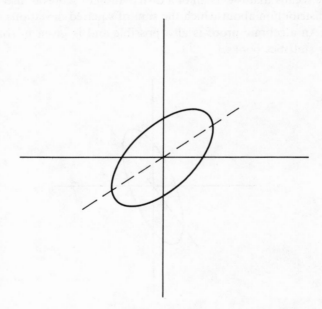

*Figure 2-8   Scatter plot of standard scores. The variances of the two variables are equal, so the ellipse is at a 45° angle to the axes. The regression line passes through the origin.*

The mathematical definition of the product moment correlation coefficient is given by the formula

$$r_{xy} = \frac{\sum\limits_{i=1}^{N} Z_{x_i} Z_{y_i}}{N},$$ (2.10)

which may be expressed in words as the mean of the cross products of standard scores. This definition does not help much in understanding what correlation is, and the associated formula is not useful for computation. However, when we recall that $Z_{x_i} = x_i/S_x$ and $Z_{y_i} = y_i/S_y$ we can substitute these values into the definitional formula, giving us

$$r = \frac{\Sigma xy}{NS_x S_y}$$ (2.11)

Now, this formula looks a little more familiar. The sum of deviation score crossproducts in the numerator is the same quantity we had in one of the formulas for the regression coefficient. The difference is that in this case the denominator contains two correctionlike terms, the two standard deviations; whereas the formula for $B$ in equation (2.8) contained only a correction for the predictor variable. This difference, of course, reflects the fact that the definition of the correlation coefficient requires equalized variances. Finally, by substituting in raw score equivalents for the various terms in the above formula we get the computing formula,

$$r = \frac{N\Sigma XY - \Sigma X \Sigma Y}{\sqrt{N\Sigma X^2 - (\Sigma X)^2} \sqrt{N\Sigma Y^2 - (\Sigma Y)^2}}$$ (2.12)

This formula, although it looks much more complicated than the definitional one, is much better for computation, because all the work is done with raw scores. The fact that this computing formula can be derived from our basic definitional formula should convince the reader of the validity of the contention we made in the first chapter, that it really does not matter whether our measurements are on interval or ratio scales, because they are converted to standard score scales in the process of computing the correlation coefficient. Standard scores have an arbitrary origin at the mean, which makes them interval scales.

The correlation coefficient has several important interpretations, some of which we discuss in detail and some which we mention but do not discuss in depth.

## SLOPE OF THE REGRESSION LINE

The first aspect of the correlation coefficient develops out of our previous discussion of the least-squares regression line. When scores are expressed in standardized form, $r$ is the slope of the best fit regression line for predicting $Y$ from $X$. We noted that the value of $B$ was the same regardless of whether raw or deviation scores were used, but that the value of $B$ changed, depending on which variable was considered to be the predictor variable, $X$. We gave

$$B = \frac{\Sigma xy}{\Sigma x^2}$$

as the definition of $B$. But, when standard scores are used the sum of squared deviations about the mean is always equal to $N$, the number of observations. Thus the value of the denominator in the formula for $B$ *using standard scores* is the same, regardless of which variable is considered the predictor. That is, $\Sigma Z_x^2 = \Sigma Z_y^2 = N$. This means that there is only one regression line for predicting $Y$ from $X$ when standard scores are used, regardless of which variable is called $Y$, and that the formula for predicting $Y$ from $X$ using $Z$ scores has the form

$$Z'_y = r_{yx} Z_x \tag{2.13}$$

By converting our measures to standard scores we may use the correlation coefficient for prediction purposes.

We will use the data from Table 2-1 to illustrate the computation of $r$ and its use in prediction. Substituting the appropriate values from Table 2-2 into the computational formula for $r$ we get

$$r = \frac{N\Sigma XY - \Sigma X \Sigma Y}{\sqrt{N\Sigma X^2 - (\Sigma X)^2} \sqrt{N\Sigma Y^2 - (\Sigma Y)^2}}$$

$$= \frac{10(107{,}123) - 677(1{,}565)}{\sqrt{10(45{,}999) - (677)^2} \sqrt{10(256{,}471) - (1{,}565)^2}}$$

$$= \frac{11,725}{\sqrt{1,661} \sqrt{115,485}} = \frac{11,725}{(40.76)(339.8)}$$

$$= \frac{11,725}{13,850.25} = .85$$

This correlation of .85 means that for every standard score unit of increase in $X$, regardless of which variable is considered to be $X$, we expect an increase of .85 standard score units in $Y$. Using the data for our hypothetical individual who is 69 inches tall and weighs 165 pounds, we find that his standard scores for height and weight, respectively, are

$$Z_H = \frac{+1.3}{4.076} = +.3189 \text{ and } Z_W = \frac{9.17}{33.98} = +.2699$$

To predict his weight from his height we have

$$Z'_W = .85(.3189) = +.2711$$

When we convert this value back into a raw score we get predicted weight $= .2711(33.98) + 156.5 = 9.21 + 156.5 = 165.71$, which differs from our previous result only by a rounding error. Note also the similarity between $Z_W$ and $Z'_W$.

The correlation coefficient bears a precise mathematical relationship to the regression coefficient which is given by the formula

$$B = r \frac{S_Y}{S_X}$$

We can see that this is true by substituting the values of the standard deviations from our height and weight example into this formula. Where weight is to be predicted from height we get

$$B = .85\left(\frac{33.98}{4.076}\right) = .85(8.337)$$

$$= 7.086$$

which differs from the value obtained in Table 2-2 only by rounding error. For the situation in which height is to be predicted from weight, we observe the following result

$$B = .85\left(\frac{4.076}{33.98}\right) = .85(.1199) = .10196$$

which is again extremely close to our previous value. The reader should note that it is extremely important to remember which variable is being considered as $X$ and which as $Y$.

From the foregoing discussion there is one important thing which becomes apparent. The only time the two regression coefficients become equal is when the variances of the two variables are equal. When this occurs, both $B$s will be equal to the correlation coefficient, because the ratio of the two variances will be unity regardless of which is in the numerator and which is in the denominator. This condition is met when the variables are expressed in standard score form. Thus the correlation coefficient is a regression coefficient in this special case, and there is only one such coefficient for expressing the relationship between two sets of scores. The reader who wishes a more mathematical discussion and proof of the above relationship should consult one of the standard statistical texts, such as McNemar (1969).

### PROPORTION OF VARIANCE

A second interpretation of correlation coefficient is closely related to $\eta^2$, the correlation ratio. As we showed earlier, $\eta^2$ is an index of the degree to which knowing a person's score on $X$ reduces our error in predicting his score on $Y$. We have also seen that the regression equation can be used for making the most accurate predictions when $X$ is continuous. However, the regression coefficient itself is only the slope of this best prediction line and does not give direct information about the proportion of variance in $Y$ that is predictable from $X$. A second interpretation of $r$ does permit direct statements about the degree of association between the two variables, or stated another way, the accuracy of prediction.

In our discussion of $\eta^2$ we defined the standard error of estimate $[S_{Y(X)}]$ as the standard deviation of the distribution of $Y$ scores within a category of $X$ about the mean of those $Y$ scores and said that this value was assumed to be representative for all the categories of $X$. Because the regression line (defined by $B$ and $A$, $B$, or $r$, depending on whether raw, deviation, or standard scores are being considered) runs through the $Y$ means of the infinitely many $X$ categories when $X$ is continuous and the relationship is linear, the standard deviation of deviations from the regression line is also a standard error of estimate. That is, the standard error of estimate is the standard deviation of the distribution of errors in prediction, regardless of whether that prediction is from category means or a regression line.

If we define an error in prediction as $Y - Y'$, then the variance of the distribution of these errors is given by the formula

$$S_{y \cdot x}^2 = \frac{\Sigma(Y - Y')^2}{N}$$

Note that we are now using the symbol $S_{y \cdot x}^2$ (read "the variance of $y$ given $x$") rather than $S_{Y(X)}^2$ (read "the variance of $y$ within $x$") to distinguish cases where prediction is made from a continuous regression line from those where prediction is from $Y$ means within a category of $X$. If the scores are expressed as deviation scores, which will not change their variance, and if we substitute $r(S_y/S_x)x$ for $y'$ we get the formula

$$S_{y \cdot x}^2 = \frac{\Sigma\left(y - r\dfrac{S_y}{S_x}x\right)}{N}$$

It is possible by simple algebra (see McNemar, 1969, p. 138) to show that this formula becomes

$$S_{y \cdot x} = S_y \sqrt{1 - r_{yx}^2} \qquad (2.14)$$

By squaring both sides of the equation and shifting terms we get

$$r_{xy}^2 = 1 - \frac{S_{y \cdot x}^2}{S_y^{\,2}}$$

which is precisely the same result we obtained with $\eta^2$. By using the same principle we used before, namely, that $S_y^{\,2} = S_{y'}^{\,2} + S_{y \cdot x}^2$, we see that

$$r_{xy}^2 = \frac{S_{y'}^{\,2}}{S_{y'}^{\,2}}$$

In other words, the square of the correlation coefficient can be viewed directly as the proportion of variance in $Y$ scores, which is predictable from $X$. This is a very useful result, because it means that we can get from one index, $r$, both an equation for prediction and an indication of how accurate that prediction is likely to be. (The equivalence between

$r^2$ and $\eta^2$ requires that the assumption of linearity of regression is met. The correlation ratio provides an index of relationship for any situation, but failure to meet the linearity assumption will cause $r^2$ to underestimate the degree of relationship. In general, we may say that $r^2$ will be less than or equal to $\eta^2$, depending on the validity of the linearity assumption.)

The fact that it is $r^2$ rather than $r$ which reveals the proportion of variance in $Y$ that is predictable from $X$ should lead the reader to be somewhat critical of comments that appear from time to time in the research literature to the effect that a correlation of .40 indicates a substantial relationship between two variables. A correlation of this magnitude indicates that only 16% of the variance in one variable is predictable from the other, a situation which becomes all the more depressing when we realize that it takes a correlation of .707 before we can claim that 50% of the variance of one variable is predictable from the other.

## ACCURACY OF PREDICTION

A look at the standard error of estimate gives even greater cause for caution. Suppose we have a variable, say, a measure of achievement, that has a standard deviation of 10 and we wish to predict achievement scores using some other variable. Table 2-3 presents the standard errors of estimate we would have if our predictor were correlated at one of

**TABLE 2-3**
*Some Typical Correlations and
the Standard Errors of
Estimate when $S_y$ is 10.*

| $r$ | $\sqrt{1 - r^2}$ | $S_{Y.X}$ |
|-----|------------------|-----------|
| .10 | .995 | 9.95 |
| .20 | .98  | 9.80 |
| .30 | .95  | 9.50 |
| .50 | .866 | 8.66 |
| .70 | .714 | 7.14 |
| .80 | .60  | 6.00 |
| .90 | .436 | 4.36 |
| .95 | .33  | 3.30 |
| .99 | .04  | .40  |

several levels with the achievement measure. Because the standard error of estimate is the standard deviation of the distribution of errors and because we assume that the errors have a normal distribution, we can draw inferences about the number of errors we would make of a given magnitude. Under the assumption of $r = 0$ or no information, 32% of our predictions will be in error by 10 units or more (32% of the normal distribution is at least 1 standard deviation from the mean). When $r = .10$, 32% of our predictions will be in error by at least 9.95 units. With an $r$ of .50 we will be in error by at least 8.66 units in about one-third of the cases. Even when the correlation is .90 we have only reduced our uncertainty to the degree that we are in error by less than 5 units in about 75% of cases. Obviously, we can hardly justify confidence in precise predictions made on the basis of correlations of .40 to .60, and this is the range of most of the validity coefficients found in the literature.

It would be inappropriate to conclude from this discussion that we should give up on attempts to study or predict behavior just because we cannot be very accurate in most cases at the present time. There are situations in which a large number of predictions are to be made and where relatively small improvements in accuracy are worthwhile. There are also situations in which only errors in one direction are important. However, it is well to keep in mind that even correlation coefficients of substantial magnitude reflect relatively modest improvements in predictive accuracy.

We may also note that the correlation coefficient is a descriptive index as well as a predictive tool. For this descriptive purpose, that is, to reflect the degree of association between two variables or the measures of two traits, correlations of all magnitudes are useful. They reflect more or less accurately, depending on the adequacy of the sampling, the degree of relationship between two variables. We must keep in mind this distinction between the correlation coefficient as a descriptive index and as a predictive tool, because the distinction is very important in some of our later discussions of multivariate techniques.

## PROPORTION OF COMMON ELEMENTS

This fourth way of interpreting the correlation coefficient is rarely seen in statistics texts, probably because there are very few situations in which such an interpretation is very close to the actual case. However, viewing $r$ as an index of the proportion of elements two groups have in common may help the reader to visualize the proportion of variance interpretation. To this end, we present McNemar's formula (9.15), which he gives without proof (1969, p. 145)

$$r_{xy} = \frac{Nc}{\sqrt{Nx + Nc}\sqrt{Ny + Nc}}$$

In this formula $Nx$ is the number of elements that are unique to the set $X$; $Ny$ is the number of elements unique to the set $Y$; and $Nc$ is the number of common elements.

Another way to view the situation described above is provided by using a Venn diagram. Venn diagrams are pictures that show different types and degrees of relationships between sets of objects, numbers, or elements of any kind. Figure 2-9 shows some Venn diagrams that depict

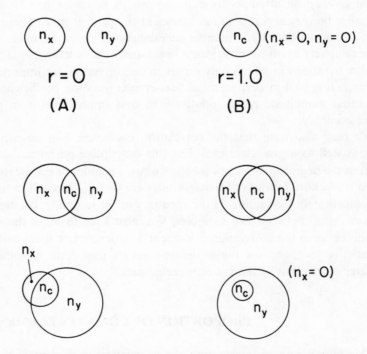

Figure 2-9    Venn diagrams showing the "proportion of common elements" interpretation of the correlation coefficient.

$$K(x:y) = \frac{Nc}{Nc + Nx} = 1 - \frac{Nx}{Nc + Nx}$$

$$r = 1 - \frac{Nx^{'}}{Nx^{'}}$$

various levels of correlation. When there are no common elements between two sets, $Nc = 0$, the circles do not overlap at all, and $r = 0$ (Figure 2-9*a*). In Figure 2-9*b* we have the case where all elements in either set are in common with the other, resulting in $Nx = Ny = 0$ ($Nc \neq 0$) and $r = 1.0$. That is, when the overlap between the two sets is complete, the correlation between them is perfect. The other diagrams in Figure 2-9 show various levels of correlation between the two extremes. Note that it is not possible to obtain a negative correlation from this formula, and, indeed, a negative correlation from this approach would be very difficult to interpret!

Some authors, most notably Cronbach et al. (1972), view variance from the perspective of Venn diagrams. This provides the essential connection between common elements and common variance for us. We have seen that $r^2$ provides us with the proportion of variance in one variable that is predictable from another. If we substitute the idea of variance for the idea of a set of elements that, within the set, differ from one another and hence vary, we can come up with a Venn diagram such as that in Figure 2-10. This diagram has one important difference from the ones in Figure 2-9, namely that the two circles are, necessarily, the same size. This is so because the two variables have been standardized, thus equating the variances of the two variables and therefore the sizes of the circles. We can see clearly from this diagram that the proportion of variance in $Y$ that is associated with $X$ is the same as the proportion of variance in $X$ that is associated with $Y$. The symmetry of this relationship is of importance to us in understanding later developments. The

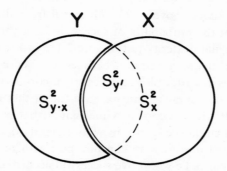

*Figure 2-10   Venn diagram showing $r^2$ as the proportion of common variance. Remember that $r^2 = S_{y'}^2/S_y^2$.*

property of symmetry does not necessarily hold for the proportion of common elements interpretation of $r$.

## EXPECTATION OF SUCCESS

The fifth interpretation of the correlation coefficient is quite useful when we are dealing with a prediction situation in which we wish to predict success or failure on some task. This is often the case in applied research where we are studying the empirical validity of an instrument for use in occupational selection. Let us say, for example, that we have a test of manual dexterity and we wish to use this test to aid in the selection of mechanics. Our criterion of success as a mechanic might be a rating of quality of work made by the supervisor. Then we might get a scatter plot like those in Figure 2-11 showing the relationship of standard score on the rating scale to standard score on the manual dexterity test.

There is a way by which we can use the information in these two scatter plots to determine the probability of success or failure of a given individual. After making the scatter plot we add up the number of individuals who obtain predictor scores in each of the given ranges on that variable and enter these in the row labeled $f$. Then we find the number of individuals in each column whose criterion scores exceed the cutoff defining success and enter this number in the row $f'$. The ratio $f'_i/f_i$ gives the probability that an individual with a predictor score $(X)$ in category $X_i$ will be successful on the job.

If we define success on the job as equivalent to obtaining a rating with a standard score above $-.75$ we can tell from the scatter plots in Figure 2-11 what the probability of success is for a person with a given score on the manual dexterity test. Figure 2-11$a$ is a scatter plot such as we might obtain if the correlation between test score and rating were .70. Part $b$ shows a scatter plot representing a correlation of .84. Let us suppose that in our organization we have found that manual dexterity is correlated .70 with success as a mechanic (that is, part $a$ applies). If a prospective employee comes in with a standard score of $-.50$ on the test of manual dexterity we can hire him with the knowledge that there is a 65% chance he will prove successful according to our criterion of success. If, however, our company applies a more stringent criterion, that success is defined as obtaining a standard score rating of greater than $+.25$, our new recruit has only an 18% chance of being judged successful. These probabilities are obtained from the figure by dividing the number of candidates in that $X$-score category who were successful (11 in the row labeled $f' > -.75$; 3 in the row labeled $f' > +.25$) by the total number of candidates scoring in that category of $X$ (17).

Now let us assume that the same test is used by another company (B) for the same purpose, but that for some reason the test has higher validity in the second company, as shown in Figure 2-11$b$. By applying the first criterion of success, company B can hire the same individual with the knowledge that he has an 82% chance of proving satisfactory. Under the second criterion, the chance of his success in company B is only 12%. Thus, regardless of which criterion of success is used, company B can be much more confident of the outcomes of its selection procedure than company A can be, because the correlation of the test with the criterion is higher.

Another way to view the problem of probability of success in correlation is to look at the number of correct decisions made from the scatter plot. A correct decision is made whenever we *either* predict success for an individual who turns out to be successful *or* predict failure for a person who does not succeed. We do this by means of setting a cutting score on the predictor variable in the scatter plot and predicting success for everyone whose predictor score is greater than the cutting score. Likewise, we predict failure for everyone below the cutting score. The *hit rate* is the proportion of correct predictions made by using a given cutting score.

Suppose we have a situation like that in Figure 2-11$a$ and that we define success on the job as obtaining a supervisor rating above $+.25$. We can choose various cutting scores to see which will give us the highest hit rate. The outcomes from using six different cutting scores are given in terms of the number of correctly predicted successes, correctly predicted failures, predicted successes who fail (false positives), and predicted failures who succeed (misses) in the top portion of Table 2-4 for the data in Figure 2-11$a$. The same information for the situation where the correlation is higher (from Figure 2-11$b$) is given in the bottom half of the table.

There are three things that are quite apparent from Table 2-4. First, as the cutting score is raised from $-.75$ to $+.25$ the hit rate rises to a maximum, then it goes down as the cutting score is further increased. Second, the hit rate is uniformly better for the scatter plot with the

Dexterity Z Scores.

| Rating Z scores | < -1.75 | -1.75 to -1.26 | -1.25 to -.76 | -.75 to -.26 | -.25 to +.25 | +.26 to +.75 | +.76 to +1.25 | +1.26 to +1.75 | > +1.75 | $f_y$ |
|---|---|---|---|---|---|---|---|---|---|---|
| > +1.75 | | | | | | 1 | 1 | 1 | 1 | 4 |
| +1.26 to +1.75 | | | | | 1 | 2 | 1 | 1 | 2 | 7 |
| +.76 to +1.25 | | | | 1 | 1 | 4 | 4 | 2 | | 12 |
| +.26 to +.75 | | | 1 | 2 | 4 | 6 | 3 | 1 | | 17 |
| -.25 to +.25 | | 1 | 2 | 3 | 8 | 2 | 2 | 1 | 1 | 20 |
| -.75 to -.26 | 1 | 1 | 3 | 5 | 3 | 2 | 1 | 1 | | 17 |
| -1.25 to -.76 | 1 | 2 | 4 | 3 | 2 | | | | | 12 |
| -1.75 to 1.26 | 1 | 2 | 1 | 2 | 1 | | | | | 7 |
| < -1.75 | 1 | 1 | 1 | 1 | | | | | | 4 |
| $f$ | 4 | 7 | 12 | 17 | 20 | 17 | 12 | 7 | 4 | |
| $f' > -.75$ | 1 | 2 | 6 | 11 | 17 | 17 | 12 | 7 | 4 | |
| $f' > +.25$ | 0 | 0 | 1 | 3 | 6 | 13 | 9 | 5 | 3 | |
| $\% > -.75$ | .25 | .29 | .50 | .65 | .85 | 1.00 | 1.00 | 1.00 | 1.00 | |
| $\% > +.25$ | 0 | 0 | .08 | .18 | .30 | .76 | .75 | .71 | .75 | |

*Figure 2-11a. Dexterity Z scores.*

| Rating Z scores | < −1.75 | −1.75 to −1.26 | −1.25 to −.76 | −.75 to −.25 | −.25 to +.25 | +.26 to +.75 | +.76 to +1.25 | +1.26 to +1.75 | > +1.75 | $f_y$ |
|---|---|---|---|---|---|---|---|---|---|---|
| > +1.75 |  |  |  |  |  |  | 1 | 1 | 2 | 4 |
| +1.26 to +1.75 |  |  |  |  | 1 | 1 | 1 | 2 | 2 | 7 |
| +.76 to +1.25 |  |  |  | 1 | 2 | 3 | 4 | 2 |  | 12 |
| +.26 to +.75 |  |  |  | 1 | 4 | 7 | 3 | 2 |  | 17 |
| −.25 to +.25 |  |  |  | 4 | 8 | 5 | 3 |  |  | 20 |
| −.75 to −.26 |  | 1 | 3 | 8 | 4 | 1 |  |  |  | 17 |
| −1.25 to −.76 | 1 | 2 | 5 | 3 | 1 |  |  |  |  | 12 |
| −1.75 to −1.26 | 1 | 3 | 3 |  |  |  |  |  |  | 7 |
| < −1.75 | 2 | 1 | 1 |  |  |  |  |  |  | 4 |
| $f_x$ | 4 | 7 | 12 | 17 | 20 | 17 | 12 | 7 | 4 |  |
| $f > -.75$ | 0 | 1 | 3 | 14 | 19 | 17 | 12 | 7 | 4 |  |
| $f > +.25$ | 0 | 0 | 0 | 2 | 7 | 11 | 9 | 7 | 4 |  |
| $\% > -.75$ | 0 | .14 | .25 | .82 | .95 | 1.00 | 1.00 | 1.00 | 1.00 |  |
| $\% > +.25$ | 0 | 0 | 0 | .12 | .35 | .65 | .75 | 1.00 | 1.00 |  |

*Figure 2-11b.   Dexterity Z scores.*

*Figure 2-11    Expectancy tables for predicting success. In part (a) r = .70. In part (b) r = .84. See text for explanation.*

higher correlation, but not much so. Third, as the cutting score rises, the relative frequency of the two types of erroneous predictions changes. This last point is important to keep in mind, because the consequences of various types of incorrect decisions may be different in practical situations. For example, if a company predicts success for a prospective employee who turns out to be a failure, the company will have lost all the time and money spent to train the person without getting the return of his labor. On the other hand, not detecting potentially satisfactory workers may be even more costly. Some other examples of the consequences of decisions are given in the context of utility theory by Cronbach and Gleser (1965).

One other feature of the selection situation may be changed to alter the hit rate (or it may be made necessary by external demands of the situation). This feature is the proportion of applicants who must be taken in in order to fill manpower requirements in an employment situation or the number of applicants who can be taken because of limitations of

**TABLE 2-4**

*Hit Rates for Various Cutting Scores for the Expectancy Tables in Figure 2-11. Success is Defined as a Rating of + .25 or Greater.*

| Cutting Score | Correctly Predicted Successes | Correctly Predicted Failures | Misses | False Positives | Hit Rate |
|---|---|---|---|---|---|
| From Figure 2-11*a* | | | | | |
| −1.25 | 40 | 11 | 0 | 49 | 51% |
| − .75 | 39 | 22 | 1 | 38 | 61% |
| − .25 | 36 | 36 | 4 | 24 | 72% |
| + .25 | 30 | 50 | 10 | 10 | 80% |
| + .75 | 17 | 54 | 23 | 6 | 71% |
| +1.25 | 8 | 57 | 32 | 3 | 65% |
| | | | | | |
| From Figure 2-11*b* | | | | | |
| −1.25 | 40 | 11 | 0 | 49 | 51% |
| − .75 | 40 | 23 | 0 | 37 | 63% |
| − .25 | 38 | 38 | 2 | 22 | 76% |
| + .25 | 31 | 51 | 9 | 9 | 82% |
| + .75 | 20 | 57 | 20 | 3 | 77% |
| +1.25 | 11 | 60 | 29 | 0 | 71% |

space, personnel, or facilities in a mental health situation. In either case, the proportion of the applicant population to be taken is called the *selection ratio*, and the value of this ratio has important implications for setting the cutting score.

Let us say, for example, that a company finds its predictor test to have the relationship with job success depicted in Figure 2-11a so that the top half of Table 2-4 applies. If we assume that the company must find 35 satisfactory employees out of 100 applicants, then a cutting score of $-.25$ must be used, even though this value does not give the highest hit rate. The manpower demands require a less-than-optimum cutting score to obtain a sufficient number of successful workers. It may even be necessary to lower the criterion of satisfactory performance to obtain enough people to fill the available spots. An illustration of what happens to the hit rates for various cutting scores when the criterion of success for our example is lowered to $-.75$ is given in Table 2-5 for the data in Figure 2-11a. Meehl and Rosen (1955) have discussed the effect of finding different frequencies of a trait (in our example, success on the job; in theirs, presence of a clinical syndrome) in the population. They conclude that predictive devices such as psychological tests will be of greatest value when the frequency of a characteristic in the population, its *base rate*, is about 50% and that tests will be essentially useless when the base rate is near zero or 100%. Tables showing the number of satisfactory employees to be expected for various selection ratios with several levels of test validity have been prepared by Taylor and Russell (1939).

**TABLE 2-5**

*Hit Rates for Various Cutting Scores when the Definition of Success is Reduced to $-.75$ Rating.*

| Cutting Score | Correctly Predicted Successes | Correctly Predicted Failures | Misses | False Positives | Hit Rate |
|---|---|---|---|---|---|
| $-1.25$ | 74 | 8 | 3 | 15 | 82% |
| $-.75$ | 68 | 14 | 9 | 9 | 82% |
| $-.25$ | 57 | 20 | 20 | 3 | 77% |
| $+.25$ | 40 | 23 | 37 | 0 | 63% |
| $+.75$ | 23 | 23 | 54 | 0 | 46% |
| $+1.25$ | 11 | 23 | 66 | 0 | 34% |

## THE NORMAL CORRELATION SURFACE

The sixth interpretation of the correlation coefficient is of more importance to the mathematical statistician than it is to the behavioral scientist who wishes to interpret a set of data. We present it here because some readers may encounter it elsewhere in the statistical literature and because it will help us later in understanding one aspect of discriminant function analysis.

One of the underlying assumptions of the product-moment correlation coefficient is that the distributions of the two variables involved in the relationship each have normal distributions in the population. Suppose we take two such variables and plot their joint frequency distribution (which is what we do when we make a scatter plot). However, instead of making tallies to represent individuals, we will represent each person by a cube. When we have completed our joint frequency distribution we will get a solid figure somewhat like a symmetrical mountain. This figure will be made of piles of little cubes representing individuals with particular combinations of scores on the two variables. Under the assumption that the two variables have normal distributions themselves, the result of their joint distribution will be what is called a bivariate normal surface, which is the bivariate analogue of the normal curve. However, there is one important difference. The shape of the bivariate normal surface depends on the degree of correlation between the two variables. If the variables are uncorrelated, the resulting surface is round, such that slicing it vertically through the center at any angle to the axes results in the same normal curve. If the correlation between the variables is nonzero, the bivariate normal surface will be elongated, either into the upper-right and lower-left quadrants if the relationship is positive or into the upper-left and lower-right quadrants for a negative correlation. The higher the correlation, the greater the degree of elongation, until at the extremes ($r = \pm 1.0$) the surface becomes a unidimensional normal curve along the regression line. (The reader should be able to see that this will happen because all observations must lie exactly on the regression line for a correlation of $\pm 1.0$ to occur.)

There is another aspect of this joint frequency distribution that may help the reader visualize it and appreciate its characteristics. If we have made the joint frequency distribution and we slice it horizontally at any given point, the *shape* of the flat surface so revealed will be constant, regardless of the level at which we slice, the only difference between levels being the size of the resulting flat surface. Also, this surface will be of the same shape as the outline of a scatter plot made in the ordinary way from an ideal random sample from the population that gave us the bivariate normal surface. Increasing sample size is equivalent to taking

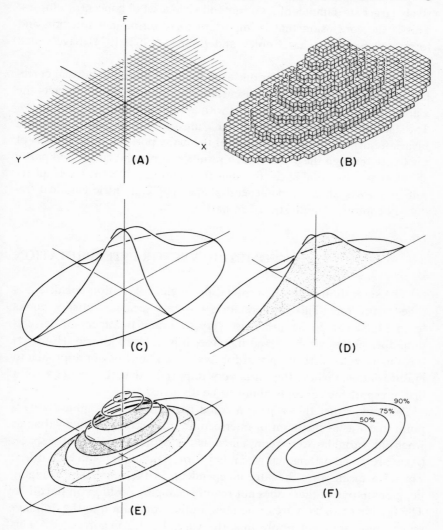

*Figure 2-12   Construction of a bivariate normal distribution. See text for explanation.*

the slice lower in the figure until, at the extreme, the bottom of the figure is the shape and size of the scatter plot for the whole population. Thus we are justified in assuming that the scatter plot from a random sample gives us a good approximation of the relationship in the population. We are also justified in making the important assumption that successively larger subsamples of a sample will yield a set of concentric ellipses. This is the conclusion that is important for discriminant function and forms the basis of what Cooley and Lohnes (1962, 1971) have called Centour analysis.

Figure 2-12 shows the various aspects of the bivariate normal correlation surface which we have discussed. Part *a* of the figure depicts the *x* and *y* axes in the horizontal plane with a vertical axis (*f*) for frequency. The stacked cubes are shown in *b*, and *c* illustrates the finished, smoothed bivariate surface. Part *d* shows one of the many possible vertical slices of the figure through the *f* axis which results in a normal curve, and part *e* shows that horizontal slices through the figure at different levels all result in ellipses of the same general shape. The concentric elliptical frequency contours are illustrated in part *f*.

## GEOMETRIC VECTOR INTERPRETATION

The seventh and final interpretation of the correlation coefficient is a geometric one. This interpretation is rarely seen in statistics books (even McNemar, in his otherwise very thorough treatment of correlation, does not mention it), and therefore it is not well known. However, it is this view to correlation that forms the backbone of our approach to multivariate statistics; thus it is very important that the reader have a firm grasp of the concepts about to be discussed.

We saw earlier that a person could be represented graphically by a point that was at the point of intersection of his scores on two variables when the variables were represented as the orthogonal axes of a space. (Axes are called orthogonal when they form right angles with each other. This also means that they are independent, in the sense that changing position on one of them does not require changing position on the other. The fact that two axes are orthogonal in this application to the ordinary scatter plot does not imply that the variables are uncorrelated.) This representation of variables as orthogonal axes permitted us to develop the idea of correlation and the applications we have seen to this point.

Now let us turn the situation around and say that the people rather than the variables form the axes of our space. The variables may then be represented as points whose locations are given by the scores people obtained on the variables. Let us say also that we are dealing with standard scores, a restriction on the data that does not cost anything in terms of the generality of the conclusions we can draw, but which does make it possible to simplify our discussion.

There is an inherent problem in trying to visualize a "reverse scatter plot" of this kind. When we had two variables and they served as the axes of our space, it was a very simple matter to show graphically what the scatter plot would look like for any number of individuals. Because we have the statistical requirement that the number of people be fairly large, we are now faced with the problem of trying to visualize a space with higher dimensionality. It is possible to visualize and to represent on paper the case in which we have three individuals (note that we can now represent any number of variables simultaneously with little difficulty), but when it comes to the situation in which we have four or more persons on whom we have data the graphic approach fails us. If the reader does his best to extend the presentation to situations in which pictures fail us, we can proceed. The reader may be confident that we are on mathematically firm ground, even though we exceed the limits of graphic representation, which we must do because it is essentially pointless to discuss a correlation where only three people have been observed.

Figure 2-13 shows how two variables might be plotted in a "people space" of three dimensions. That is, we have shown our people as axes 1, 2, and 3 and our two variables as points $A$ and $B$. The point marking variable $A$ is shown as though individual 1 has a score of $+.5$, individual 2 has a score of $-.75$, and individual 3 has a score of $+2.0$. The scores giving the coordinates of the point representing variable $B$ are $-1.2$, $+.6$, and $-.3$, respectively, for the same three individuals. We should note that the situation depicted in Figure 2-13 is for a subgroup of three from a larger sample. That this must be the case is seen from the fact that the sum of standard scores for the group on which they have been computed must be zero. There must have been individuals in the group

who had negative scores on *A* and positive scores on *B* to balance out
those shown in the figure. A subset is used here to simplify the explanation while making it possible to show the process of plotting. Point *C*
in Figure 2-13, which represents another variable and shows that we
can handle several variables simultaneously by forming a scatter plot in
this way, is included to demonstrate that the fact that the standard
scores sum to zero does not mean that the point so defined will fall at
the origin (or place where the axes intersect). The coordinates of this
point are $+0.8$, $+1.2$, and $-2$, respectively. This situation generalizes
to the case in which we have a large group of people whose scores on the
variables have been standardized.

At this point it is useful to present a technical term which will reappear
periodically throughout the remainder of this book and to consider two

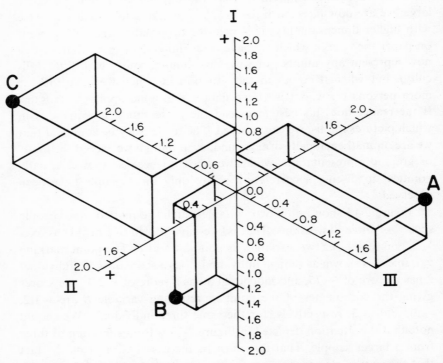

*Figure 2-13   Plotting 3 variables in a space where the axes are people. Note that
this is a standard score space and that the 3 people have been chosen from a larger
group.*

somewhat different but related concepts to which the term applies. This technical term is the word *vector*, and it has two meanings that are important for our discussion. The first of these is a concept from matrix algebra. A vector is a linear array of numbers. For our purposes we can think of the set of standard scores obtained by a group of people on a test as a vector. Such a set of scores is generally set out in the following way:

| Individual | Score on variable X |
|:---:|:---:|
| 1 | $X_1$ |
| 2 | $X_2$ |
| 3 | $X_3$ |
| ⋮ | ⋮ |
| $i$ | $X_i$ |
| ⋮ | ⋮ |
| $N$ | $X_N$ |

and we may speak of this array as the vector **X** composed of the elements or individual scores $X_i$. When expressed vertically, as above, the set forms a *column vector*. The same set of scores could be arranged horizontally to form a *row vector* without changing the meaning of the numbers.

The second meaning of the word vector comes from geometry and physics. A vector is a line in space that has direction and length. In this use, the word comes to refer to an amount of something, a something that might be force, as in physics, or test score, as in psychology. A vector of this kind is usually represented by an arrow going from a point of origin to a point of termination. This arrow represents the resultant effect from a combination of sources of that effect. The location of the point of the arrow is defined by the coordinates of the amounts of the various effects involved.

We are now in a position to put these two definitions of vector to use. First, we may note that our array of scores for a group of people on a variable, be it an ability test, a learning task, or a personality inventory, can be expressed as a vector in the first sense. Thus we can speak of a vector of scores and know that we are referring to the array of scores obtained by a group of individuals. The vector can be composed of raw scores, deviation scores, or standard scores, depending on the use to which the scores are to be put. Second, we may note that our vector of scores gives us the coordinates of the point of termination of a vector in the second sense of the term. That is, the vector of scores on the variable defines the vector (or arrow) representing the variable in a space defined by the people. This arrow or *variable vector* runs from the origin to the point representing the variable and has the same meaning as the point itself. Figure 2-14 represents the vectors resulting from the three people's scores on the three variables in Figure 2-13.

Representing variables as vectors in a space of people has some rather useful consequences for us. For example, there is a theorem in geometry that can be extended to prove that when coordinates are expressed as deviations from a mean, the length of the vector defined by those coordinates is given by the sum of the squared coordinates. Putting this statement into a mathematical expression, we have vector length $= \Sigma x^2$ for deviation scores, which may be recognized as the definition of the variance of the variable being represented by the vector times $N$, the number of individuals. Thus the length of a variable vector in a

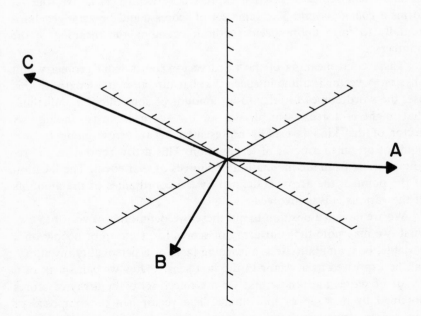

*Figure 2-14    Variables represented as vectors for the plot in Figure 2-13.*

space of people is proportional to the variance of the variable. When we represent variables in terms of a standard score coordinate system of people, each vector is of the same length, and that length is $N$ times the variance of standard scores, or $N$.

A second useful result of the foregoing developments also comes from a geometric theorem. If we take what is called the inner product of two deviation score vectors (that is, the sum of crossproducts of equivalent scores) and divide by $NS_xS_y$ we get the correlation between the two variables. Specifically, if we have two deviation score vectors for $X$ and $Y$,

| Individual | Deviation Score on $X$ | Deviation Score on $Y$ |
|:---:|:---:|:---:|
| 1 | $x_1$ | $y_1$ |
| 2 | $x_2$ | $y_2$ |
| 3 | $x_3$ | $y_3$ |
| ⋮ | ⋮ | ⋮ |
| $i$ | $x_i$ | $y_i$ |
| ⋮ | ⋮ | ⋮ |
| $N$ | $x_N$ | $y_N$ |

then the inner product is given by multiplying $x_1$ times $y_1$, $x_2$ times $y_2$, and so on and summing for all $N$ scores. Performing the operations described above gives us the expression $\Sigma xy/NS_xS_y$ which will be recognized as equation (2.11), one of the definitional formulas for the correlation coefficient. Obviously, if the score vectors are expressed in standard score form the expression simplifies to $\Sigma Z_x Z_y/N$.

However, the really important development that comes from treating variables as vectors in space comes from the fact that either of the above expressions also defines *the cosine of the angle between the two vectors*. That is, the cosine of the angle between two variable vectors in a standard score space defined by $N$ individuals (a space of $N$ dimensions) is equal to the mean of the products of the coordinates defining those vectors, which is equal to the correlation between the two variables. This gives us a means of graphically representing the correlation between two or more variables, which is very useful later on. (A formal proof that the correlation between two variables is equal to the cosine of the angle between the variable vectors in an $N$-dimensional space of people is given in Chapter 4 of Harman, 1967.)

Consideration of what it means to have variables expressed in standard score form and represented geometrically yields some interesting results. When two variables have a high positive correlation, we generally think about them as measuring the same or highly similar things. Geometrically, a high correlation is shown by two lines being close together in space, with a small angle between them. As an angle approaches zero, its cosine approaches 1.0, which, by our recent development, translates into a correlation approaching 1.0. Thinking about scores in standardized form, we see that the only time a relationship between two variables can be perfect in the sense of a correlation of 1.0 is when each person's $Z_x$ is equal to his $Z_y$. When this condition is met (i.e., $Z_{x_i} = Z_{y_i}$),

$$\frac{\Sigma Z_x Z_y}{N} = \frac{\Sigma Z_x{}^2}{N} = \frac{\Sigma Z_y{}^2}{N} = 1.0$$

But when $Z_x = Z_y$ for all individuals, the two variable vectors have the same coordinates and therefore fall one on top of the other, yielding an angle of zero which has a cosine of 1.

A perfect negative relationship can only occur when $Z_x = -Z_y$ for each individual. In a case such as this

$$\frac{\Sigma Z_x Z_y}{N} = -\frac{\Sigma Z_x{}^2}{N} = -\frac{\Sigma Z_y{}^2}{N} = -1.0$$

and we can see geometrically that the points defined by these coordinates fall in exactly opposite positions with respect to the origin, yielding an angle of 180° and a cosine of −1.0. This geometrically reflects the fact that we usually think of two variables with a high negative correlation as measuring essentially opposite traits. By way of general summary, it may be said that a positive correlation will result in an acute angle between the variable vectors, a negative correlation gives an obtuse angle, and zero correlation or independence yields a right angle (which has a cosine of 0).

To fully appreciate the close relationship between correlation as it is manifest in the scatter plot of people, with variables as the axes, and the geometric approach to correlation in which we have a scatter plot of variables and people are the axes, we will make a brief excursion into a multivariate situation. Suppose that we have measured $N$ people on $m$ variables and we wish to express the relationships among the variables by a scatter plot. The data can conveniently be presented in the form of a *matrix* (which is a collection of score vectors) as shown below.

$$
\text{Matrix } \mathbf{X} =
\begin{array}{cccc}
\begin{array}{c}\text{Variable}\\\text{Vector}\\x_1\end{array} &
\begin{array}{c}\text{Variable}\\\text{Vector}\\x_2\end{array} &
\begin{array}{c}\text{Variable}\\\text{Vector}\\x_3\end{array} &
\begin{array}{c}\text{Variable}\\\text{Vector}\\x_m\end{array}\\[4pt]
\begin{bmatrix}x_{11}\\x_{12}\\x_{13}\\\vdots\\x_{1i}\\\vdots\\x_{1N}\end{bmatrix} &
\begin{bmatrix}x_{21}\\x_{22}\\x_{23}\\\vdots\\x_{2i}\\\vdots\\x_{2N}\end{bmatrix} &
\begin{bmatrix}x_{31}\\x_{32}\\\vdots\\\vdots\\\vdots\\\vdots\\x_{3N}\end{bmatrix} & \cdots &
\begin{bmatrix}x_{m1}\\x_{m2}\\\vdots\\x_{mi}\\\vdots\\x_{mN}\end{bmatrix}
\end{array}
$$

$$
=
\begin{array}{l}
\text{Individual vector 1}\\
\text{Individual vector 2}\\
\\
\\
\\
= \text{Individual vector } i\\
\\
\\
\\
\text{Individual vector } N
\end{array}
\begin{bmatrix}
[x_{11}x_{21}\cdots x_{m1}]\\
[x_{12} \quad\cdots\quad ]\\
\vdots \quad \cdots \quad \vdots\\
\vdots \quad \cdots \quad \vdots\\
\vdots \quad \cdots \quad \vdots\\
[x_{1i} \quad \cdots x_{mi}]\\
\vdots \quad \cdots \quad \vdots\\
\vdots \quad \cdots \quad \vdots\\
\vdots \quad \cdots \quad \vdots\\
[x_{1N} \quad \cdots x_{mN}]
\end{bmatrix}
$$

To represent all these data in a single scatter plot *of people*, we would use a space of $m$ dimensions, with the $m$ variables serving as the coordinates of the space. Each person would be located by the point defined by his scores on the $m$ variables. Note that our previous discussion of this type of scatter plot treated the special case where $m = 2$. The bivariate scatter plot for any pair of variables can be generated by considering only the appropriate two columns of the matrix. For example, the scatter plot (in the conventional sense) showing the relationship between variables $x_1$ and $x_3$ would be prepared by plotting points with coordinates $(x_{11}, x_{31})$, $(x_{12}, x_{32}) \cdots (x_{1i}, x_{3i}) \cdots (x_{1N}, x_{3N})$ to give the points representing the $N$ people. If we wish to form a scatter plot of variables, we would use a space of $N$ dimensions defined by the people and each variable would be located in the manner outlined earlier in this section. That is, to plot the $m$ variables in the space defined by two people (say, individual 2 and individual $i$), we would use the two row vectors for these individuals. Each variable would then be plotted as a point with coordinates $(x_{12}, x_{1i})$ for variable 1, $(x_{22}, x_{2i})$ for variable 2, and so forth. Of course, in all practical applications it would be necessary to use the full "people space" of $N$ dimensions, but to do so takes us beyond our limits for visualization.

The upshot of viewing a set of $m$ scores for each of $N$ people as a matrix is that it shows that the two ways of making a scatter plot are nothing more or less than two different ways of representing the same information. To form a scatter plot of people, we consider each column vector of the matrix to be a dimension, with the scores in that column representing the values of the $N$ people on that dimension. To form a scatter plot of variables, we view each row vector of the matrix as a dimension, with scores in the row representing values of the $m$ variables on that dimension. The two scatter plots presented earlier are merely special cases of this general principle. Any set of data can be presented in either form without loss of information. The only practical difference is that one representation may present the clearest picture for one interpretation of correlation, whereas the other is better for a second.

## SUMMARY

In this chapter we have developed the logic and a little of the mathematics of the product-moment correlation between two variables. A least-squares strategy, in which the object is to minimize the sum of squared differences between predicted and actual scores, provides the basic rationale for correlation and regression. Using the criterion that a system for prediction should be a least-squares solution with the information at hand, we saw that the overall mean of $Y$ scores is our best prediction for each individual when no information is available. When group membership on a nominally scaled $X$ variable is available, the mean $Y$ of the individual's subgroup is the least-squares best prediction, and $\eta^2$ (eta squared) is an index of the proportion of variance in $Y$ that is associated with $X$. As $X$ becomes a continuous variable and the relationship between $X$ and $Y$ is constrained to fit a linear model, the $Y$ means of the infinitely many $X$ groups come to form a straight line, the regression line, which is given by the equation $Y' = BX + A$ and which provides a least-squares prediction of $Y$. By adding the additional restriction that $X$ and $Y$ be expressed as standard scores, we found that a single index, the correlation coefficient $r$, defined the least-squares regression line.

After developing $r$, we explored some of its interpretations. We saw that $r$ gives the slope of the best fit regression line for standard scores and that this coefficient is the same regardless of which variable is considered to be the predictor (a feature which does not hold for the regression coefficient). The square of $r$ can be interpreted as the proportion of variance in $Y$ which is predictable from (or associated with) $X$ and the standard error of estimate $(= Sy\sqrt{1 - r^2})$ is the standard deviation of errors of prediction or an index of the spread of scores around the regression line. The correlation coefficient can also be interpreted as the proportion of elements two sets have in common or the amount of overlap in the two variables when they are expressed in a Venn diagram. A somewhat indirect application of $r$ is its use in making statements about the probability of success of a given individual.

We also saw that the correlation coefficient can be used to help describe the shape of a bivariate normal distribution and that the scatter plot of people for the two variables is given by taking a horizontal slice across the bivariate normal figure. Finally, we explored the geometric or vector interpretation of correlation and saw that the correlation coefficient is equal to the cosine of the angle of separation of two variable vectors plotted in a space defined by the $N$ people in the sample. This last interpretation gives us a simplified means of viewing the complex relationships among many variables in a multivariate analysis.

# 3

## OTHER INDICES OF RELATIONSHIP FOR TWO VARIABLES

In the previous chapter we developed the product moment correlation coefficient (pmc) as an index of the strength of relationship between two variables when both variables were continuous and the relationship was assumed to be linear. We also saw that $\eta^2$ gave an indication of the strength of relationship when one variable was continuous and the other was nominal. This index could be used with a relationship of any form and could be applied to a continuous variable when a nonlinear relationship was found simply by forming categories of the $X$ variable. In this chapter we present methods of correlation that may be used when, for various reasons, a product moment coefficient is inappropriate. Some of the indices to be discussed are themselves special cases of the pmc for use with particular kinds of data; others are estimates of what the pmc would be if the data had certain properties.

## SPECIAL PRODUCT MOMENT CORRELATIONS

**Rank order correlation.** When we went from $\eta^2$ to the regression line we required that our $X$ variable be a continuous intervally scaled variable, rather than achieving only a nominal level of scaling. The reader probably noticed that we skipped over the ordinal level of scaling. We did this for two reasons. First, the move from nominal to interval scaling for the predictor variable provides the most direct route from the correlation ratio to the pmc. The criterion or predicted variable was already assumed to be scaled so that it had interval properties, and any consideration of the ordinal case for $X$ would have led us astray. Second, there is no convenient index of relationship for an ordinal $X$ and an interval $Y$ except that given by forming groups on $X$ and computing $\eta^2$.

When data are ordinal they are most often in the form of ranks, and if they are in some other form they can be converted to ranks quite easily. Many, perhaps most, of the data collected by behavioral scientists are of an ordinal level of scaling and, as such, are not appropriate for analysis using a pmc of the ordinary kind (even though the authors of most journal articles go ahead and analyze them as though they had interval properties). When one or both of the variables to be analyzed have only ordinal properties we are on the safest ground if we use an index of relationship that makes only ordinal assumptions. Such an index is given by the Spearman rank order correlation coefficient, $\rho$(rho), which is a pmc between two sets of data that are in the form of ranks.

The regular pmc is inappropriate for ordinal data, because several of its assumptions are not fulfilled. For example, it is senseless to talk about ordinal data having a normal distribution or a linear relationship, because the units of measurement are of unequal and probably unknown size. For the same reason it is not appropriate to talk about standard scores with ordinal data, because the mean and standard deviation are meaningless. We are thus faced with the necessity of finding some index of relationship that does not require the concepts of mean and variance. The index we need is given by the formula

$$\rho = 1 - \frac{6\Sigma d^2}{N(N^2 - 1)} \tag{3.1}$$

which was derived by Spearman in 1902. In this formula the term $\Sigma d^2$ refers to the sum of squared differences between the ranks for $N$ people on two variables. The logic and computations are shown in Table 3-1, in which data are shown for 10 people on two hypothetical rating scales. In this table we have shown the scores obtained by the 10 people, the conversion of these scores into ranks, the computation of the $d^2$s, and computation of $\rho$.

The layout of Table 3-1 shows that each person's score on a variable is converted into a rank, the rank of 1 being given to the score indicating the greatest amount of the trait. These ranks could perfectly well be reversed and 1 assigned to the score indicating the least amount of the trait without changing the results, so long as the procedure was consistently applied to both variables. Care must be taken at this step that the problems of interpreting directionality of measurement, discussed in Chapter 1, do not occur. Note that tied scores, such as the $Y$ scores for individuals 4 and 5, are handled by assigning the same rank to both scores. The rank assigned is the average of the ranks in question (in this case the ranks 9 and 10, whose average is 9.5). Tied ranks do not otherwise effect the computational procedures.

**TABLE 3-1**

*Computation of the Rank Order Correlation Coefficient (rho) for Data from 10 Hypothetical People.*

|    | X Score | Y Score | X Rank | Y Rank | $d = (X - Y)$ | $d^2$ |
|----|---------|---------|--------|--------|---------------|-------|
| 1  | 42      | 130     | 4      | 7      | $-3$          | 9     |
| 2  | 36      | 133     | 8      | 5      | 3             | 9     |
| 3  | 44      | 140     | 2      | 2      | 0             | 0     |
| 4  | 45      | 119     | 1      | 9.5    | $-8.5$        | 72.675 |
| 5  | 32      | 119     | 9      | 9.5    | $-.5$         | .25   |
| 6  | 37      | 135     | 7      | 3      | 4             | 16    |
| 7  | 27      | 123     | 10     | 8      | 2             | 4     |
| 8  | 41      | 134     | 5      | 4      | 1             | 1     |
| 9  | 40      | 132     | 6      | 6      | 0             | 0     |
| 10 | 43      | 142     | 3      | 1      | 2             | 4     |
|    |         |         |        |        | $\Sigma d^2 =$ | 115.925 |

$$\rho = 1 - \frac{6\Sigma d^2}{N(N^2 - 1)} = 1 - \frac{6(115.925)}{10(100 - 1)} = 1 - \frac{695.55}{990} = 1 - .703 = .297$$

The rank order correlation is a pmc between the two sets of ranks (i.e., if we used the ranks from columns 3 and 4 of Table 3-1 in the computing formula for $r$, we would get the same value we obtain from Table 3-1), but its interpretation is much more restricted than that of a pmc computed in the regular way on interval data. Although there are seven different ways in which we could interpret $r$, there is only one permissable interpretation of $\rho$ as the amount or degree of relationship between the two variables. It cannot be interpreted as the slope of regression line, because no regression line can be found. Neither can it be used to compute a standard error of estimate, nor any of the other uses of the pmc. If a set of rank order correlations are to be used as a basis for a multivariate analysis, the results should be interpreted with extreme caution.

Determining the statistical significance of $\rho$ presents something of a problem, because the statistic is usually used with small samples. The appropriate test for the significance of $\rho$ (the probability of whether the obtained $\rho$ is a chance deviation from a population value of zero) is given by the formula.

$$t = \rho \sqrt{\frac{N - 2}{1 - \rho^2}}$$

in which $t$ has $N - 2$ degrees of freedom. However, this distribution is an approximation to the $t$ distribution, and when sample size is less than 10 the approximation is none too good. We are thus in a bind. We would generally use $\rho$ with small samples, because applying it to large samples often results in many tied ranks, reducing the effectiveness of our ranking procedure. But when sample size becomes too small, the adequacy of our test of significance deteriorates. McNemar (1969) mentions the possibility of using one of Kendall's tau statistics (to be discussed later in this chapter) in place of $\rho$ when $N$ is small. By way of illustration, the test of significance for the data in Table 3-1 yields

$$t = .297 \sqrt{\frac{10 - 2}{1 - .297^2}} = .297 \sqrt{\frac{8}{.912}} = .297 \sqrt{8.77} = .879$$

which is not statistically significant with $df = 8$. Rho has the same limits of $\pm 1.0$ that $r$ has (which are reached when the ranks show perfect agreement or perfect disagreement), and a $\rho$ of zero indicates that no relation is present. Also, $\rho$ can be computed in any situation in which

$r$ can be justified. However, the latter is the statistic of choice when its assumptions are met because of the wider range of interpretations which are permissible.

**Point biserial correlation.** Occasionally, situations arise in behavioral science research in which it is necessary to find the degree of relationship between an $X$ variable that is dichotomous and a $Y$ variable that is on an interval scale. This situation can occur in two ways, one in which the dichotomous variable is a true dichotomy and another in which the variable is considered to be a measure of a continuously varying trait, but the scores are reported dichotomously either for convenience or because a more refined measure is not possible. The only common example of a variable of the first type is sex; one is either male or female (at least in most cases). Multiple choice test items provide a common example of a variable that is measured dichotomously (an item is either right or wrong) but which is assumed to be a reflection of an underlying continuous trait. There are different special correlation coefficients for each of these two cases. At this point we now discuss the coefficient for the relationship between a truly dichotomous variable and a continuous variable, because this index *is* a pmc. We reserve discussion of the index of relationship for a dichotomously measured continuous variable for the next section, because this coefficient is an estimate of a pmc.

The coefficient that gives the relationship between a truly dichotomous and a continuous variable is called the *point biserial correlation coefficient*. Its value is given by the formula

$$r_{pb} = \frac{(\overline{Y}_2 - \overline{Y}_1)\sqrt{p_1 p_2}}{S_y} = \frac{\overline{Y}_1 - \overline{Y}_T}{S_y}\sqrt{\frac{p_2}{p_1}}$$

The meanings of the terms in this equation are as follows:

$\overline{Y}_1 = $ the mean of the $Y$s for those people in the first category of $X$
$\overline{Y}_2 = $ the mean of the $Y$s for those people in the second category of $X$
$\overline{Y}_T = $ the mean of all $Y$s
$p_1 = $ the proportion of people who are in the first $X$ category
$p_2 = $ the proportion of people who are in the second $X$ category
$S_y = $ the standard deviation of all $Y$s around $\overline{Y}_T$.

The value obtained for $r_{pb}$ in the above equation is the same as the value that would have been obtained if $X$ had been scored 0 and 1 (or 2 and 3, or any other numerical values) and an ordinary pmc had been computed in the usual way. The only real reason for the existence of this alternate formula is that it is computationally much simpler when a computer is not available than is the pmc formula, which requires crossproducts. (Rho is also a computational shortcut.)

The point biserial correlation is obviously closely related to the $t$ test for the significance of the difference between two means, involving as it does the difference between the means divided by $N - 1$ times the standard error of the mean and an adjustment for relative sample sizes. In fact, for small samples ($N$ less than 30) the $t$ test for the difference between the two means ($\overline{Y}_1$ and $\overline{Y}_2$) provides a test of the statistical significance of $r_{pb}$. For samples greater than 30, the value $1/\sqrt{N}$ may be taken as the standard error of $r_{pb}$ and the test of significance has the form

$$Z = \frac{r_{pb}}{1/\sqrt{N}}$$

Although $r_{pb}$ is a pmc, it behaves somewhat differently from the $r$ computed for two continuous variables. When there is no relationship between the two variables, $r_{pb}$ will be exactly zero, a situation that can occur only when $\overline{Y}_1 = \overline{Y}_2$. (If the reader thinks back to our opening remarks about using means for least squares prediction, he will see that only when the group means are equal is no information conveyed or predictive accuracy gained by knowing a person's group membership.) However, although $r$ has a range from $+1.0$ to $-1.0$, $r_{pb}$ has limits of $\pm.798$, and these limits can be reached only when $p_1 = p_2 = .50$ and there is no overlap between the two groups. When there is overlap between the groups or the sample sizes are unequal ($p_1 \neq p_2$), $r_{pb}$ will be less than the above value. In the extreme case in which one group includes 90% of the subjects, the limits of $r_{pb}$ are about $\pm.58$, even when there is no overlap between one group and the other. This restriction in effective range must be kept in mind when interpreting the point biserial correlation. Notwithstanding this feature, $r_{pb}$ is the appropriate coefficient to use when a truly dichotomous variable is correlated with a continuous variable, and $r_{pb}$s can be used in combination with regular $r$s when some type of multivariate analysis is to be performed on the data.

We can illustrate the computation of $r_{pb}$ using the data in Table 3-2. Here we have six males and four females who have the scores shown on the test (whatever it is). The $r_{pb}$ is .63, indicating a fairly strong tendency for the members of the higher-numbered group, the females, to obtain higher test scores. Note that referring to either group as the higher-numbered group is arbitrary, but if we had designated the males as group 2, the correlation would have been $-.63$. Because the values of $X$ did not enter into the computations, we could have chosen any numerical (or verbal) labels we wished for the groups.

The point biserial correlation can be interpreted in exactly the same way as the rank order coefficient. That is, although it is a product moment correlation, it is only an index of degree of relationship and does not support the other interpretations listed for $r$.

**Phi.** Because there are relatively few truly dichotomous variables in behavioral science, there are even fewer situations in which we are likely to need an index of relationship between two truly dichotomous variables. However, should such a situation arise, the appropriate coefficient to

**TABLE 3-2**

*Computation of the Point Biserial Correlation Between Sex and Test Score.*

| $X$ (sex, 1 = male) | $Y$ (test score) | |
|---|---|---|
| 1 | 25 | $\bar{Y}_1 = \dfrac{138}{6} = 23.0$ |
| 2 | 28 | |
| 1 | 24 | |
| 1 | 22 | $\bar{Y}_2 = \dfrac{107}{4} = 26.75$ |
| 2 | 24 | |
| 2 | 30 | |
| 1 | 26 | $\bar{Y}_T = \dfrac{245}{10} = 24.5$ |
| 1 | 21 | |
| 2 | 25 | $p_1 = .6 \qquad p_2 = .4$ |
| 1 | 20 | |

$$S_y = \sqrt{\frac{10(6087) - (245)^2}{100}} = \sqrt{8.45} = 2.91$$

$$r_{pb} = \frac{(26.75 - 23)\sqrt{.6(.4)}}{2.91} = \frac{3.75(.49)}{2.91} = \frac{1.8375}{2.91} = .63$$

use is $\phi$(phi). This statistic reflects the degree of relationship between two truly dichotomous variables such as sex and voting behavior. (Note that we are talking about casting one's vote for one of two candidates or for or against some issue, not the attitudes or beliefs underlying the behavior. In the latter case, one of the variables would be a dichotomized continuum.) Phi is also called the fourfold point correlation, because its use applies to a fourfold frequency table (a $2 \times 2$ table).

If we collect data on sex and voting behavior, the data might be cast into a fourfold table of the form

Vote

|         | For | Against |           |
|---------|-----|---------|-----------|
| Males   | $A$ | $B$     | $A + B$   |
| Females | $C$ | $D$     | $C + D$   |
|         | $A + C$ | $B + D$ | $A + B + C + D = N$ |

in which cell $A$ contains the number of males who voted for the issue, cell $B$ the number of males voting against the issue, and cells $C$ and $D$ contain the same information for females. $A + B$ is the total number of males, $C + D$ is the total number of females, $A + C$ is the total number of positive votes and $B + D$ is the total number of negative votes. When the data have been prepared in this way the formula

$$\phi = \frac{BC - AD}{\sqrt{(A + B)(C + D)(A + C)(B + D)}}$$

gives the correlation between sex and voting behavior. A positive correlation would indicate a tendency for males to vote against the issue and females to vote for it, because only when this is true will $BC$ be greater than $AD$. Of course, a negative correlation would be interpreted as indicating the reverse situation. The statistical significance of $\phi$ may be tested by

$$\chi^2 = N\phi$$

in which $\chi^2$ has 1 degree of freedom, or by

$$Z = \frac{\phi}{1/\sqrt{N}}$$

when $N$ is large (say, greater than 100). An example of the computation of $\phi$ using some hypothetical data for the situation described above is given in Table 3-3. Here we find a correlation of $+.503$, which yields a $\chi^2$ of 50.3. This $\chi^2$ is significant at the .01-level, indicating that there is probably a nonzero relationship between sex and vote, with men tending to cast negative votes and women positive votes. ($Z = 5.03$, indicating also that the relationship is nonzero.)

Since $\phi$ is a *product moment correlation,* it can be used with point biserial and regular correlation coefficients when performing more complex analyses. However, because the direction of scaling of both variables is completely arbitrary, care is necessary in its interpretation. It is subject to the same interpretive limitations as $r_{pb}$, but, unlike the point biserial correlation, $\phi$ can reach values of $\pm 1.0$ when all the cases fall in either the $A$ and $D$ or the $B$ and $C$ cells. When $\phi$ is less than unity, its maximum value will be affected by whether the marginal frequencies $A + B$, $A + C$, $B + D$ and $C + D$ are equal. The more these marginals deviate from equality, the lower the maximum possible correlation, which is exactly what we saw for the proportions $p_1$ and $p_2$ in the case of $r_{pb}$. We have more to say about the matter of marginals later.

**TABLE 3-3**

*Computation of Phi Between Sex and Voting.*

|  | Votes | |  |
|---|---|---|---|
|  | For | Against |  |
| Males | 10 | 40 | 50 |
| Females | 35 | 15 | 50 |
|  | 45 | 55 | 100 |

$$\phi = \frac{BC - AD}{\sqrt{(A + B)(C + D)(A + C)(B + D)}} = \frac{40(35) - 10(15)}{\sqrt{(50)(50)(55)(45)}}$$

$$= \frac{1400 - 150}{\sqrt{6187500}} = \frac{1250}{2487} = .503$$

$$\chi^2 = N\phi = 100(.503) = 50.3$$

## ESTIMATES OF PRODUCT MOMENT CORRELATIONS

There are two senses in which we can say that an index of relationship is an estimate. In one sense, all correlation coefficients, whether they are pmc's, such as those we have discussed to this point, or are not pmc's, such as the coefficients we discuss in the remainder of this chapter, are estimates of the degree of relationship between two variables in a population based on data from a sample from that population. In this sense we have been dealing with statistical estimation in which tests of significance are appropriate. However, in this section we are dealing with indices that are not themselves pmc's but are best guesses about what the pmc in the *sample* would be if our data were of a different form. The differences, then, between the coefficients discussed in this section and those of the previous section are two. These coefficients are based on assumptions about the nature of the traits underlying the variables that are not reflected in the way in which the data are reported and scaled, and, the values obtained from these coefficients cannot be obtained by applying the usual formula for the pmc to the data at hand. More specifically, the two coefficients discussed below are estimates of what the pmc would have been if the data, which are scaled as dichotomies, had been measured on a continuous scale.

**Biserial correlation.** There are many occasions, particularly in psychological measurement, where the traits underlying both the variables can reasonably be assumed to vary continuously, but where one of the variables has been scaled as a dichotomy, either out of necessity or for convenience. This situation must be clearly distinguished from the case in which one variable is truly dichotomous and the other is continuous, the situation in which the point biserial correlation is appropriate. In this new case in which the dichotomized variable is assumed to be really continuous and the dichotomy is artificially imposed, the proper statistic to indicate the degree of relationship between the two variables is the biserial correlation coefficient, $r_b$, which is given by the formula

$$r_b = \frac{(\overline{Y}_2 - \overline{Y}_1)(p_1 p_2)}{h S_y} = \frac{(\overline{Y}_2 - \overline{Y}_T) p_2}{h S_y}$$

in which all terms have the same meaning they had for computing the point biserial correlation and the new term, $h$, is the height of the normal curve (its ordinate) above the base line for $p_1$ (or $p_2$, whichever is smaller).

The value of $h$ can be found from most tables of the normal curve by finding the row in which the proportion of the curve beyond the tabled $Z$ value is equal to the smaller of $p_1$ or $p_2$ and looking across to the column headed "ordinate" or "$h$." Many of the widely used statistics texts (e.g., McNemar, 1969; Guilford and Fruchter, 1973) give tables that can be used for this purpose.

The use of the normal curve table in computing the biserial $r$ implies one of the assumptions underlying the statistic, that the variable represented as a dichotomy is in fact continuous and normally distributed in the population. One situation in which the biserial $r$ is commonly used and in which this assumption is likely to be met is in correlating items with test scores. This may involve what is called a part-whole correlation, in which the dichotomously scored item is part of the total test or scale with which it is to be correlated, or it may be used to correlate an item, such as expression of opinion for or against a particular issue, with a test or scale of which it is not a part. There is some difference of opinion about whether it is proper to use part-whole correlations, but the problem is not too serious unless the tests have very few items. In the first place, many older studies used part-whole correlations, thus their use can be defended on the basis of comparability with previous research. Secondly, modern computer technology has made it a relatively simple matter to rescore the test so that a total score with each item omitted in turn can be obtained. This removes the problem completely and is probably the best course to take in future work. The reason given for not using part-whole correlations is that the presence in the total score of the item being studied tends to make the correlation slightly higher than it should be, because that item contributes identical variance to both variables, all of the variance on one side of the relationship and part of the variance on the other.

A place in which the part-whole problem may go unrecognized but in which it is potentially much more serious is in the correlation of one subscale from a large instrument with another subscale from the same instrument (or another instrument that uses some of the items of the first). The Minnesota Multiphasic Personality Inventory (MMPI) is a good example of an instrument where this may happen. Many of the items on the MMPI appear on more than one scale. On those scales on which the common items are keyed in the same direction, they tend to make the correlations among the scales higher than they would be if they were composed of nonoverlapping sets of items, whereas for the scales that share items keyed in opposite directions the correlations tend to be spuriously lowered. These same biases will, of course, appear when the MMPI scales are correlated with other instruments that have borrowed items from the MMPI. The only solution with existing instruments is to exercise caution in the interpretation of such coefficients, because eliminating items from the total scores will change whatever construct and content validity have been built up for the scales.

In Table 3-4 we present some hypothetical data that show the computational procedures involved in $r_b$. The computations in the table show that $r_b$ is .80 regardless of which form of the formula is used. However, these values show us something else about $r_b$, because the data in Table 3-4 are the same as those in Table 3-2, with the exception that the $X$ variable has been called "item score" instead of "sex" and it has been coded 0 for wrong, 1 for correct, instead of 1 for male, 2 for female. But, the value of the point biserial correlation for these same data was .63, not the .80 we find in Table 3-4. These two coefficients point up a nearly

**TABLE 3-4**

*Computation of the Biserial Correlation Between Item Score and Total Test Score.*

| $X$<br>(item score,<br>1 = correct) | $Y$<br>(test<br>score) | |
|:---:|:---:|:---|
| 0 | 25 | $\overline{Y}_0 = 23.0$  $\overline{Y}_1 = 26.75$ |
| 1 | 28 | |
| 0 | 24 | $\overline{Y}_T = 24.5$ |
| 0 | 22 | |
| 1 | 24 | $p_0 = .6$  $p_1 = .4$  $h = .385$ |
| 1 | 30 | |
| 0 | 26 | $S_y = 2.91$ |
| 1 | 25 | $r_b = \dfrac{(\overline{Y}_1 - \overline{Y}_0)(p_0 p_1)}{hS_y} = \dfrac{(26.75 - 23.0)(.6)(.4)}{.385(2.91)} = \dfrac{.90}{1.12} = .80$ |
| 0 | 20 | |

or

$$r_b = \frac{(\overline{Y}_1 - \overline{Y}_T)p_1}{hS_y} = \frac{(26.75 - 24.5)(.4)}{.385(2.91)} = \frac{2.25(.4)}{1.12} = \frac{.9}{1.12} = .80$$

universal fact about the relationship between $r_b$ and $r_{pb}$. With the one exception of a correlation of zero, the absolute value (disregarding sign) of $r_b$ will be greater than that of $r_{pb}$. Actually, the relationship between the two coefficients can be shown to be

$$r_b = \frac{\sqrt{p_1 p_2}}{h}(r_{pb})$$

Because the value of $\sqrt{p_1 p_2}/h$ is always greater than 1.0, $r_b$ will always be greater than $r_{pb}$.

When we were discussing $r_{pb}$, we saw that the coefficient had its maximum possible value when $p_1 = p_2$ and that this maximum, which never equalled 1.0, dropped off as the inequality between the $ps$ became more extreme. The biserial correlation presents us with some more vexing problems, the most important of which is that sometimes a set of data will yield a correlation greater than unity. McNemar (1969) suggests that this is most likely to occur when the distribution of the continuous variable is decidedly not normal. Because this situation is contrary to the limits of the pmc, we must conclude that our assumption of the normality of the data has been violated. A second problem associated with the use of $r_b$ is that the sampling error of the statistic, which is used for estimating statistical significance and is given by the formula

$$S_{r_b} = \frac{\dfrac{\sqrt{p_1 p_2}}{h} - r_b}{\sqrt{N}}$$

is an approximate sampling error. The value of $S_{r_b}$ depends on the degree of inequality between $p_1$ and $p_2$, being greater for greater degrees of inequality.

Unlike $r_{pb}$, the biserial correlation can reach a value of $\pm 1.0$ when its assumptions are met. This occurs when there is no overlap in the two $Y$ distributions, regardless of the degree of inequality between $p_1$ and $p_2$. Otherwise, however, $r_b$ is subject to the same restrictions of interpretation that apply to $r_{pb}$. In addition, the fact that $r_b$ is not a pmc but an estimate of what the pmc would be if we could measure the dichotomous variable on a continuum may cause some problems if we attempt to use these estimates in a multivariate analysis. Some types of factor analysis, for example, can be used with pmc's of any kind but may break down completely if estimates of pmc's (either $r_b$ or the tetrachoric correlation discussed below) are used. That is, it may be impossible to perform some necessary calculations without obtaining imaginary values when estimates of pmc's are used in place of actual product moment correlations. This situation is potentially so serious that some writers (e.g., Nunnally, 1970) advise against ever using these estimates. This position is a little extreme, particularly when the bivariate relationships are the terminal statistics (i.e., no additional analyses are to be performed), and we can be reasonably sure that our assumptions are met, as is the case with many psychological variables. However, researchers must be aware that these coefficients are subject to restrictions that do not plague pmcs.

**Tetrachoric correlation.** We have seen that $r_b$ is an index of relationship that may be used in place of the point biserial correlation when a con-

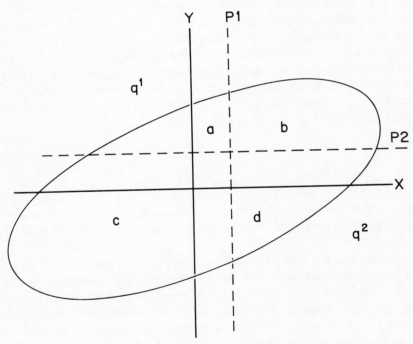

*Figure 3-1   Theoretical partitioning of the scatter plot using the tetrachoric correlation to estimate the product moment correlation. Solid lines are the axes and dashed lines are partitions. Labels a–d refer to frequencies that result from the partitions.*

tinuous variable is assumed to underlie a dichotomous measure. The tetrachoric correlation coefficient, $r_t$, bears the same relationship to $\phi$. That is, $r_t$ provides an estimate of what the pmc between two continuous variables would be when both variables have been measured as dichotomies. We might logically assume such a situation to exist when we wish to find the correlation between two ability items scored right or wrong or when we have two attitude or personality items to which or subjects have responded "agree-disagree" or "like-dislike."

One way to view the situation with which we are dealing when we use $r_t$ is to imagine that we have a fourfold table which is the result of dichotomizing both of the variables in an ordinary scatter plot as shown in Figure 3-1. The only pieces of information at our disposal are the four cell frequencies or proportions shown in Table 3-5, and our task is to estimate to the best of our ability (make our best guess about) the shape of the scatter plot that gave us those frequencies or proportions. The symbols have the following meanings:

$p_1$ and $p_2$ = proportion passing item $X$ and $Y$, respectively
$q_1$ and $q_2$ = proportion failing item $X$ and $Y$, respectively
$a$ = proportion passing $Y$ but not $X$
$b$ = proportion passing both
$c$ = proportion failing both
$d$ = proportion passing $X$ but not $Y$.

The calculation of $r_t$ is extremely complex and time consuming. Various tables and figures have been prepared to ease the labor, for example, those by Pearson (1931), and the formula given by McNemar (1969, p. 222).

**TABLE 3-5**
*Data Layout for Computation of a Tetrachoric Correlation.*

|  |  | Item $X$ | | |
|---|---|---|---|---|
|  |  | Fail | Pass | |
| Item $Y$ | Pass | $a$ | $b$ | $p_2$ |
|  | Fail | $c$ | $d$ | $q_2$ |
|  |  | $q_1$ | $p_1$ | |

The use of the tetrachoric correlation is based on the assumption that the variables underlying the fourfold frequencies would yield a normal correlation surface if they were measured on continuous scales. Given this assumption, each possible combination of frequencies or proportions in a table such as Table 3-5 would give rise to a unique scatter plot of the type shown in Figure 3-1. The combination of marginal proportions (*p*s and *q*s) and cell proportions (*a*s, *b*s, *c*s, and *d*s) would determine the shape of the ellipse and the locations of the two secondary axes shown by the dotted lines. With the shape of the scatter plot uniquely determined, the pmc describing that scatter plot is also uniquely determined.

Because it is an estimate of the pmc in the same sense that $r_b$ is, and because it may result in undefined values, the tetrachoric correlation must be viewed with the same reserve as $r_b$. It, too, may cause a multivariate analysis to break down, because computations yield imaginary values. However, like $r_b$, the tetrachoric correlation has advantages over its product moment partner, $\phi$. When its assumptions are fulfilled, it gives a good estimate of the pmc. It can reach unity and, unlike $\phi$, its value does not depend as heavily on equality of the marginal proportions. In virtually all cases $r_t$ will be greater than $\phi$, the exceptions being when both are zero ($a = b = c = d$) and when one of the cells of the fourfold table contains a 0, in which case, $r_t$ is undefined. However, because of the difficulties of computation, the restrictive nature of the assumptions, and the limits on interpretation (which are the same as those of $r_b$), $r_t$ is seldom used today.

Perhaps the most prominant use of tetrachoric correlations in the past was in Thurstone's (1938) study of the primary mental abilities (PMA). Thurstone used $r_t$ as a shortcut estimate of the pmc among his variables by forming artificial dichotomies of continuous variables and using computing diagrams to get the tetrachoric correlations, a procedure that was much quicker than computing pmc's with the facilities then available. However, this expedient has had some unfortunate consequences in recent years in that the PMA data have become a standard demonstration example in factor analysis, but the fact that they are in the form of $r_t$s makes them unanalyzable by some of the newer methods of factor analysis. This is unfortunate, because the wealth of experience that has accumulated with these data cannot be applied to the analysis of the more restrictive factoring procedures. The fault rests with the limitations of the tetrachoric correlation. If for no other reason than this, its use should generally be avoided.

## OTHER INDICES

We close this chapter by mentioning briefly other indices of relationship for use with two variables. One of these is an order statistic which, like $\rho$, concerns the agreement between two sets of ranks, the other is a measure that can be used with nominal variables. These statistics are more frequently used in sociology than in psychology and are generally inappropriate for multivariate analysis; however, they can be of benefit in some situations, and we present them as examples for the reader to consider. There are many other correlationlike statistics, of which these are a sample.

**Kendall's tau** ($\tau$) is an index of similarity between two sets of ranks in which the crucial feature is the number of inversions in ranking, rather than the difference between ranks used in computing $\rho$. The difference between the two can be seen if we recast the data from Table 3-1 into the new form of Table 3-6. Here the 10 individuals have been placed in rank order on the first variable such that person 4 was ranked first, and so on. The same procedure was carried out using the ranks on $Y$,

**TABLE 3-6**
*Ranking of Two Variables for Computing Kendall's Tau.*

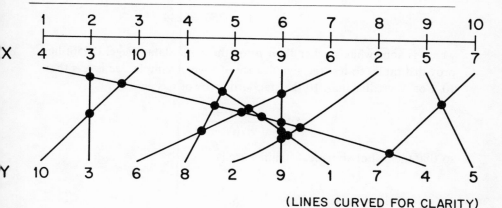

(LINES CURVED FOR CLARITY)

with the exception that the tie between 4 and 5 for last place was arbitrarily decided in favor of 4, which is necessary because $\tau$ does not handle ties well. Next, straight lines are drawn from each individual's position on $X$ to his position on $Y$. Each place where two such lines intersect is counted as an inversion in ranking. In the table each intersection is noted by a dot, of which there are 17. The value of $\tau$ is given by the formula

$$\tau = 1 - \left[ \frac{2\ (\text{number of inversions in ranking})}{\text{number of } possible \text{ pairs of objects}} \right]$$

The numerator in this ratio is twice the number of intersections (dots) in Table 3-6, in this case $2 \times 17 = 34$. The denominator is the number of possible unique pairs that can be formed from $N$ objects (subjects), taken in pairs. The general formula for combinations,

$$_rC_N = \frac{N!}{r!(N-r)!}$$

gives for the present example

$$_2C_{10} = \frac{10!}{2!(8)!} = \frac{10 \times 9 \times 8 \times 7 \times 6 \times 5 \times 4 \times 3 \times 2 \times 1}{(2 \times 1)(8 \times 7 \times 6 \times 5 \times 4 \times 3 \times 2 \times 1)}$$

$$= \frac{10 \times 9}{2 \times 1} = \frac{90}{2} = 45$$

Therefore, the value of $\tau$ for the rank-ordered data of Table 3-6 is

$$\tau = 1 - \frac{34}{45} = 1 - .756 = .244$$

which is somewhat smaller than $\rho$ for the same data. Siegel (1956) has provided tables to test the significance of $\tau$ when sample size is less than 10. For $N$ greater than 10 the standard error of $\tau$ is

$$S_\tau = \sqrt{\frac{2(2N+5)}{9N(N-1)}}$$

so that the test of statistical significance is

$$Z = \frac{\tau}{S_\tau}$$

In most cases $\tau$ is used only when sample size is small, because diagrams such as that in Table 3-6, which provide the easiest means for determining the number of inversions, become extremely complex when $N$ is at all large and also because ties, which make special adjustments to the formula necessary, become much more likely with larger $N$.

The values of $\tau$ and $\rho$, although not in exact agreement for a particular set of data such as ours, do give essentially the same information, that is, the relative agreement between two sets of ranks. Tau has the advantages of ease of computation and a more adequate statistical test when $N$ is less than 10, whereas $\rho$ has the advantages of being able to handle ties easily, which becomes a common problem with $N$ greater than 10, and of being a product moment correlation.

**The contingency coefficient** $(C)$ is essentially a $\chi^2$ and, as such, can be used with nominal data. It is an index of degree of relationship between unordered categories, but unlike any of the other indices we have discussed, the contingency coefficient is constrained to be positive. This

**TABLE 3-7**

*Computation of the Contingency Coefficient.*

| | | | | |
|---|---|---|---|---|
| 10 (30)[a] | 20 (30) | 30 (20) | 40 (20) | 100 |
| 80 (60) | 70 (60) | 25 (40) | 25 (40) | 200 |
| 160 (90) | 50 (90) | 45 (60) | 45 (60) | 300 |
| 50 (120) | 160 (120) | 100 (80) | 90 (80) | 400 |
| 300 | 300 | 200 | 200 | 1000 |

($Y$ labels the row dimension.)

$$\chi^2 = \sum \frac{(O - E)^2}{E} = 199.72$$

$$C = \sqrt{\frac{199.72}{1199.72}} = \sqrt{.1665} = .408$$

[a] Values in parentheses are expected cell frequencies.

would seem reasonable, because the data themselves are not assumed to have directionality. For a given set of data such as those in Table 3-7 we can compute

$$\chi^2 = \sum \frac{(0 - E)^2}{E}$$

which is the usual way of computing $\chi^2$. $C$ is given by

$$C = \sqrt{\frac{\chi^2}{N + \chi^2}}$$

which can have a value of zero for any size table, but which can never reach 1.0 (as the number of cells in the table becomes large, the upper limit of $C$ approaches but never reaches unity). For the data in Table 3-7 the value of $C$ is .408, which indicates some relationship between the variables.

It is rather difficult to compare the contingency coefficient with any of the other indices of relationship we have discussed, because it deals with data of a very different nature. In fact, $C$ is not an estimate of $r$ and is unrelated to any estimate of $r$. McNemar (1969, p. 230) suggests that if it is desired to estimate $r$ from data that are represented in categories such as those appropriate for computing $C$, it is first necessary to make assumptions regarding the nature of the variable underlying the categories. These assumptions are the same ones necessary for use of the tetrachoric correlation, and it is, therefore, best to reduce the contingency table to a 2 × 2 and determine $r_t$, rather than go through the procedures necessary to adjust $C$ so that it is an estimate of $r$. He also suggests that an appropriate test of the statistical significance of $C$ is given by testing the significance of the $\chi^2$ used in its computation. Both these suggestions seem reasonable. However, the contingency coefficient cannot be considered an appropriate statistic for use in any of the multivariate procedures to be discussed in later portions of this book.

### SUMMARY

In this chapter we have discussed seven indices of the relationship between two variables that may be used in special circumstances. The point biserial correlation is a pmc between a continuous variable and a dichotomous variable. The biserial correlation is an estimate of what the pmc would be if the dichotomous variable were actually continuously measured, but it is not itself a pmc. Phi is a product moment correlation between two dichotomous variables, and the tetrachoric correlation is an estimate of what the pmc would be if both dichotomies were measured on continua. Like $r_b$, $r_t$ is not a pmc, and we noted that both are of limited usefulness for multivariate applications. The rank order correlation, $\rho$, is a pmc between two sets of ranks and is appropriate for use with ordinal data. Kendall's $\tau$ is also a rank order correlation that agrees closely with $\rho$ but is not a pmc. It is particularly useful with small sample sizes. The contingency coefficient is an index of relationship for use with nominal data. Each of these coefficients has its usefulness, but some have more restrictive assumptions or greater limitations on their use than others.

# 4

# TERMINOLOGY OF MATRICES AND GEOMETRY

Although this book is designed to explain the concepts of multivariate correlational statistics with a minimum emphasis on mathematical background, there are certain mathematical terms and ways of viewing things that the reader must understand to grasp some of more complex analysis procedures. Also, these mathematical terms provide a nice, precise shorthand for conveying particular concepts that recur with great frequency in multivariate analysis. Therefore, this short chapter is devoted to developing a small, but necessary, working mathematical vocabulary. A more extended discussion of these topics may be found in Horst (1963).

## VECTOR

As we saw earlier, in our discussion of pmc, the term vector has two different but complementary meanings. A vector may be thought of as either a row or a column of numbers. When we use the term vector to refer to a row or column of numbers, the separate numbers in the vector

**95**

are called elements of the vector. In most behavioral science work a particular vector refers to measurements of one characteristic on several objects or one type of summary statistic for several characteristics or groups of subjects. For example, if we have 10 boys on whom we have obtained a measure of aggressiveness and we identify the boys by the numbers 1 to 10, we could place their aggressiveness scores in the same order and have a vector such as

| *Subject* | *Aggressiveness* |
|:-:|:-:|
| 1 | 20 |
| 2 | 31 |
| 3 | 18 |
| 4 | 22 |
| 5 | 25 |
| 6 | 37 |
| 7 | 24 |
| 8 | 30 |
| 9 | 22 |
| 10 | 17 |

The order of the boys may be arbitrary, but once that order has been formed, the positions of their scores in the vector of scores is fixed. The result is a vector of aggressiveness scores.

The real gain in using vectors to represent sets of scores is that it provides a ready shorthand for further use of the numbers, particularly when we are dealing with more than one set of such numbers. If we use the letter $X$ to refer to the variable of aggressiveness, then we can say that **x** refers to the vector of aggressiveness scores. By using a subscript, say, $i$, we can identify the score of any subject as $X_i$, an element of **x**. These same ideas apply when a vector is an array of summary statistics, say, means, for a number of characteristics or groups. $\bar{X}_i$ would refer to the mean for characteristic $X$ of group $i$.

A second use of the term vector was developed in our discussion of the seventh interpretation of the correlation coefficient. A vector, in the sense that the term is used by physicists, is a line in space with direction and length. We return to this geometric use of the term shortly, but now we must continue on our mathematical train of thought.

**MATRIX**

*A matrix is a group of vectors, all of which have the same subscript dimension.*

Suppose that we collected information on height, weight, age, and number of siblings as well as on agressiveness for the sample of 10 boys discussed above. We now have five different vectors, one for each variable, each with 10 elements. If we keep the order of the subjects constant from one vector to another, then in each vector a particular value of the subscript refers to the same subject. By placing the vectors together as shown below and using the subscript $j$ to refer to the different (column) vectors, we get a simple way of expressing a particular individual's score on a particular characteristic or variable. We may call the matrix of data $X$. Each unique element of the matrix is expressed as $X_{ij}$, where $i$ refers to the row in which the element is found (i.e., the individual to whom the score belongs) and $j$ refers to its column (which variable it represents). Thus $X_{31}$ refers to the aggressiveness of subject number 3 and $X_{85}$ refers to the number of siblings of boy number 8. Each element of the matrix can be uniquely identified by a combination of subscripts.

| Subject | Aggressiveness $(X_1)$ | Height $(X_2)$ | Weight $(X_3)$ | Age $(X_4)$ | Sibs $(X_5)$ |
|---------|------------------------|----------------|----------------|-------------|--------------|
| 1 | 20 | 62 | 110 | 10 | 1 |
| 2 | 31 | 58 | 112 | 9 | 0 |
| 3 | 18 | 54 | 95 | 11 | 2 |
| 4 | 22 | 59 | 90 | 10 | 2 |
| 5 | 25 | 60 | 105 | 11 | 3 |
| 6 | 37 | 55 | 100 | 8 | 1 |
| 7 | 24 | 63 | 115 | 10 | 3 |
| 8 | 30 | 51 | 95 | 8 | 1 |
| 9 | 22 | 53 | 103 | 9 | 1 |
| 10 | 17 | 58 | 108 | 10 | 0 |

$= X$

More complicated matrices, which are formed by collecting several simple matrices, also can be handled readily using subscript notation. Suppose we had the same five variables measured for three groups of 10 boys, each group representing a different socioeconomic level. The resulting matrix would be a solid figure with subscript dimensions $i, j,$ and $k$ of the form

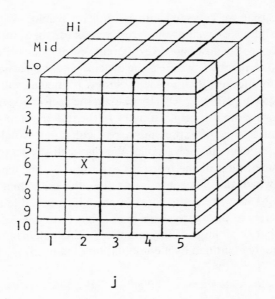

$$j$$

The score of the sixth boy in the low socioeconomic group on the height variable (the $X$ in the diagram) would be given by $X_{621}$. Any value in the matrix could be specified by giving the appropriate combination of subscripts of the general form $X_{ijk}$.

Most of the matrices with which it is necessary to operate in multivariate analysis are of the simpler kind, having only two subscript dimensions. However, the matrices of multivariate statistics are generally the result of some arithmetic operations carried out on the original groupings of scores. When we speak of a *score matrix*, we are referring to a matrix in which the rows represent individuals and the columns represent variables. The elements of a score matrix are the scores of the several subjects on the several variables. The matrix on page 97 is a score matrix, or, more specifically, a *raw score matrix*, because the elements of the matrix are raw scores. When the elements of the matrix are deviation or standard scores, we say that we have a deviation or standard score matrix respectively.

## MATRIX TERMINOLOGY

There are several terms used to describe matrices that are useful for the reader to know to facilitate the understanding of matrix operations.

The *order* of a matrix is its size in a very direct sense. A matrix that has 10 rows and 5 columns is of order $10 \times 5$ (10 by 5). In general, we can refer to any matrix as being of order $n \times m$, where $n$ is the number of rows and $m$ is the number of columns. When $n = m$ we say that we have a *square matrix*.

The *transpose* of matrix **X** (symbolized **X′**) is a new matrix that has the columns of **X** as its rows (or the rows of **X** as its columns). In schematic form, if

$$\mathbf{X} = \begin{bmatrix} x_{11} & x_{12} \\ x_{21} & x_{22} \\ x_{31} & x_{32} \end{bmatrix} \quad \text{then} \quad \mathbf{X'} = \begin{bmatrix} x_{11} & x_{21} & x_{31} \\ x_{12} & x_{22} & x_{32} \end{bmatrix}$$

If the order of **X** is $n \times m$, then the order of **X′** is $m \times n$.

A type of matrix that occurs very frequently in multivariate statistics is the *symmetric matrix*, of which the correlation matrix is a good example. A symmetric matrix is a square matrix in which the elements on one side of the *main diagonal* of the matrix are the same as those on the other side. In the diagram below we used $a_i$ to represent the elements of the main diagonal of the matrix and $b$, $c$, $d$, and so on to represent the other, *off-diagonal elements*. If we were to fold the matrix along its main diagonal,

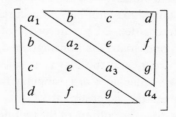

each of the off-diagonal elements in the upper triangle of the matrix would fall on the same lettered (and numerically equal) element in the lower triangle.

Another term that occurs fairly frequently when the matrix aspects of multivariate statistics are being discussed is the *determinant of the matrix*. Although determinants have several uses, there is one interpretation that is of paramount importance for us in understanding the application of matrix theory in statistics. The very important information the determinant of a matrix provides is the number of independent sources of information in the matrix being analyzed.

To see how this works, suppose that we have a matrix of correlations among 10 ability measures. This will be a symmetric matrix of order $10 \times 10$. If each of the ability measures provides some unique information that is not provided by any of the other measures, then the determinant of the $10 \times 10$ matrix will be some nonzero number. However, if one of the variables is composed entirely of information that is present in one or more of the other variables (i.e., it is linearly dependent on the other variables), the determinant of the matrix will be zero. If the redundant variable (or variables, in the case of some matrices) is removed by crossing out its row and column, the determinant of the smaller matrix will not be zero. Most multivariate statistical procedures require that the matrices being analyzed have nonzero determinants.

When the determinant of a matrix is zero, the matrix is said to be *singular*. Singular matrices are relatively rare in the behavioral sciences, but when they do occur it is necessary to find and eliminate the redundant variable or variables before a proper solution can be found. A more common problem is a nearly singular matrix, a matrix whose determinant is almost zero. A singular matrix, which can be detected by a value of zero for the determinant in the computer printout, may make solution of the multivariate equations impossible. A near-singular matrix, which is detectable by a very small value for the determinant, may permit a solution, but the values obtained may not be very accurate.

The *rank* of a matrix is defined by the order of its largest nonzero determinant. If the original matrix is nonsingular, then the rank of the matrix is equal to its order, and the matrix is said to be of full rank. If the original matrix of order $m \times m$ is singular but there is only one redundant variable, then the determinant of the $(m - 1) \times (m - 1)$ reduced matrix with the row and column of the redundant variable deleted will be nonzero and the rank of the $m \times m$ original matrix is $m - 1$.

There is really no way to determine ahead of time what the rank of a matrix will be unless variables that are known to be redundant are included in the matrix. One will be guaranteed singularity if one of the variables is computed from two or more others (e.g., variable $c = a + b$), but this is not often the case in practice. If a matrix is found to be singular, one must delete each row and its associated column in turn until a nonzero determinant is found. Each of these matrices of order $(m - 1) \times (m - 1)$ is called a *first-order minor* of the original matrix. Should the determinants of all first-order minors be zero, then there is double dependence, and the determinants of all second-order minors, found by deleting pairs of rows and columns, must be evaluated. This process continues until some minor of the matrix is found whose determinant is not zero. The rank of the matrix is the order of this largest minor with nonzero determinant.

The concept of the rank of a matrix is quite important in understanding and interpreting multivariate statistical results, because the rank of the matrix indicates the number of independent sources of variation that are present in the data. This insight into the meaning of the rank of a matrix was one of the major contributions of L. L. Thurstone in his development of multiple factor analysis in the 1930s.

## MATRIX MANIPULATIONS

A knowledge of the algebra of matrices, although not central to understanding multivariate statistics as presented in this book, may help the reader to gain a better feel for how the solutions to matrix problems are actually obtained. The most important aspects of this section are the ideas of matrix inversion and the identity matrix.

*Addition and subtraction of matrices* can be performed when the matrices are of the same order and each is done on an element-by-element basis, as shown in the following examples. If

$$\mathbf{A} = \begin{bmatrix} a_{11} & a_{12} \\ a_{21} & a_{22} \\ a_{31} & a_{32} \end{bmatrix} \quad \text{and} \quad \mathbf{B} = \begin{bmatrix} b_{11} & b_{12} \\ b_{21} & b_{22} \\ b_{31} & b_{32} \end{bmatrix}$$

then

$$\mathbf{A} + \mathbf{B} = \begin{bmatrix} a_{11} + b_{11} & a_{12} + b_{12} \\ a_{21} + b_{21} & a_{22} + b_{22} \\ a_{31} + b_{31} & a_{32} + b_{32} \end{bmatrix}$$

and

$$\mathbf{A} - \mathbf{B} = \begin{bmatrix} a_{11} - b_{11} & a_{12} - b_{12} \\ a_{21} - b_{21} & a_{22} - b_{22} \\ a_{31} - b_{31} & a_{32} - b_{32} \end{bmatrix}$$

Note that the two matrices must be of the same order if addition or subtraction operations are to be performed. In this case both matrices are $3 \times 2$. The sum or difference matrix will be of the same order as the two original matrices.

*Multiplication of matrices* is somewhat more complex. To multiply two matrices the number of columns of the first matrix must equal the number of rows of the second; the product matrix will have as many rows as the first matrix and as many columns as the second. This is necessary because matrix multiplication is performed by multiplying the elements of a row of the first matrix by the elements of a column of the second matrix and adding the products, as shown in the following example:

$$\mathbf{C} = \begin{bmatrix} c_{11} & c_{12} & c_{13} \\ c_{21} & c_{22} & c_{23} \end{bmatrix} \qquad \mathbf{D} = \begin{bmatrix} d_{11} & d_{12} \\ d_{21} & d_{22} \\ d_{31} & d_{32} \end{bmatrix}$$

$$\mathbf{C} \times \mathbf{D} = \begin{bmatrix} c_{11}d_{11} + c_{12}d_{21} + c_{13}d_{31} & c_{11}d_{12} + c_{12}d_{22} + c_{13}d_{32} \\ c_{21}d_{11} + c_{22}d_{21} + c_{23}d_{31} & c_{21}d_{12} + c_{22}d_{22} + c_{23}d_{32} \end{bmatrix}$$

Note that $\mathbf{C}$, which is of order $2 \times 3$, multiplied by $\mathbf{D}$ of order $3 \times 2$, yields a product matrix of order $2 \times 2$.

We may state a general rule that can cover all matrix multiplications. Defining matrix $\mathbf{M}$ as being of order $a \times b$ and matrix $\mathbf{N}$ as being of order $c \times d$, we can set up the multiplication as

$$\mathbf{M} \quad \cdot \quad \mathbf{N}$$
$$(a \times b) \quad (c \times d)$$

For us to be able to multiply these matrices it is necessary that the inner dimensions ($b$ and $c$) be equal. The product matrix

$$\mathbf{P} = \mathbf{MN}$$

will have $a$ rows and $d$ columns. Thus if we multiply a $10 \times 6$ matrix by a $6 \times 3$ matrix, the resulting matrix will be of order $10 \times 3$.

The process of transposing matrices plays an important part in matrix algebra, in that it makes addition, subtraction, and multiplication of matrices possible in some cases in which it would not be so without transposition. For example, matrices $\mathbf{C}$ and $\mathbf{D}$ cannot be added or subtracted in the form given above. However, by transposing either one, addition and subtraction become possible, because the order of $\mathbf{D}'$ is $2 \times 3$, which is the order of $\mathbf{C}$, and the order of $\mathbf{C}'$ is $3 \times 2$, which is the order of **D**. Thus although **C** − **D** and **C** + **D** are not solvable, the expressions **C** − **D′**, **C′** − **D**, **C** + **D′**, and **C′** + **D** can be solved.

The same thing occurs in matrix multiplication. Although it is not possible to multiply **A** × **B**, it is possible to find **A′** × **B**(= **A′B**) or **AB′**.

The reader should notice an important difference here between ordinary arithmetic and matrix arithmetic. In ordinary arithmetic the order of multiplication does not matter. The product of 8 × 6 is the same as the product of 6 × 8. In matrix arithmetic this is not true. The product of **C** × **D** in the above example was a 2 × 2 matrix. In general, **A′B** does not equal **AB′**, nor does **A′B** result in the same product matrix we would get from **BA′** (assuming that such a product is possible). Our general rule indicates that **A′B** will be of order 2 × 2, whereas **AB′** will be of order 3 × 3. Also, although both **AB′** and **BA′** are of order 3 × 3, they are not equal. In fact, **AB′** is the transpose of **BA′**, indicating still further that the order of multiplication of matrices does make a difference. There is only one exception to this principle, when both **A** and **B** are symmetric matrices so that **A** = **A′** and **B** = **B′**. Because the order of multiplication is so important, we use special terms to indicate which matrix comes first. In the expression **AB′** we say that **B′** is *premultiplied* by **A** or that **A** is *postmultiplied* by **B′**. The order in which matrices are multiplied, which is always from left to right in a matrix equation, is of great importance in determining the results of solving the equation. Note also that **C′** − **D** does not yield the same result as we would get from **C** − **D′**, although both are permissable operations.

There is one other feature of matrix manipulation that must be considered before we turn to some geometric properties and definitions. This is the idea of the *inverse of a matrix*.

In matrix algebra there is no procedure that is equivalent to division in ordinary arithmetic. However, there is a special matrix called the *identity matrix* (symbolized by **I**) that can be used for somewhat the same purpose. The identity matrix is a matrix that has ones in all elements of the main diagonal and zeros everywhere else.

$$\mathbf{I} = \begin{bmatrix} 1 & 0 & 0 & 0 & 0 & 0 \\ 0 & 1 & 0 & 0 & 0 & 0 \\ 0 & 0 & 1 & 0 & 0 & 0 \\ 0 & 0 & 0 & 1 & 0 & 0 \\ 0 & 0 & 0 & 0 & 1 & 0 \\ 0 & 0 & 0 & 0 & 0 & 1 \end{bmatrix}$$

The inverse of a matrix **A** is defined as another matrix (symbolized $A^{-1}$) that when multiplied by **A** yields an identity matrix. More specifically,

$$AA^{-1} = A^{-1}A = I$$

This equality implies that we will be finding inverses of symmetric matrices and that these inverses will themselves be symmetric matrices. When it is necessary to find the inverse of a nonsymmetric (generally, a nonsquare) matrix, $AA^{-1}$ will yield an **I** of one order and $A^{-1}A$ will yield an **I** of a different order. These orders can be found by applying the general rule discussed above. A singular matrix does not have an inverse, and when a matrix is nearly singular it is very difficult to find an exact inverse. Thus when the determinant of a matrix is very small, indicating near-singularity, the accuracy of the inverse of that matrix suffers. Because matrix inversion is required for the solution of virtually all multivariate statistical equations, it is very important to the accuracy of the results that the matrices not be singular or nearly so.

Matrix inversion is an extremely time consuming and laborious procedure, the labor of which increases rapidly as the number of variables increases beyond four or five. It is for this reason more than any other that the widespread use of powerful multivariate procedures had to await the development of high-speed computers. It is assumed throughout this book that the reader will have access to the necessary computer programs for performing the analyses described in the sections to follow.

We have given more detailed descriptions of matrix operations and terms than are really necessary to follow the development of most multivariate procedures because some feel for matrix ideas is necessary to grasp the meaning of the important concepts of matrix rank, singularity, and the inverse of a matrix. These terms will assume some more practical meaning in connection with internal factor analysis. We will be confronted with matrices of various shapes and sizes in connection with most multivariate procedures. However, the primary vehicle for understanding the several varieties of multivariate analysis is the geometry associated with matrices, to which we now turn.

## GEOMETRIC CONSIDERATIONS

Probably the most difficult skill for the person who attempts to master the geometry of multivariate analysis is the ability to visualize a space of more than three dimensions. There is no ready physical parallel to this situation, which is the reason why most discussions of multivariate procedures are couched in matrix terms. However, many people, the author included, are more comfortable with relatively concrete graphic representations than with those of an abstract mathematical nature. Welcome to the realms of higher-order geometry.

One of the advantages of a geometric approach to multivariate statistics is that it is generally possible to provide two-dimensional representations of more complex phenomena by deciding to consider only pairs of dimensions, one pair at a time. This requires selectively disregarding some information, but after some practice it can be mastered. For example, we can take a three-dimensional object such as the rectangle shown here, and by taking each pair of dimensions in turn we can get a

two-dimensional representation. The three pairs of dimensions: height and breadth, height and depth, and breadth and depth, yield these three figures

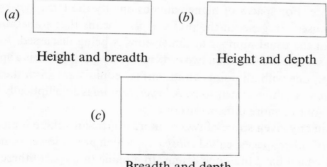

(*a*) Height and breadth    (*b*) Height and depth

(*c*) Breadth and depth

which are formed by disregarding first depth, then breadth, and finally height. Or, viewed in another and more useful way, we may think of the rectangular figure above as a stick figure, perhaps made out of wire. If we were to take a flashlight and shine it on the figure from the front (thereby omitting the depth dimension) we would get a figure like *a* above projected on a screen behind the object. If we then moved the flashlight to the side to omit the breadth dimension, the figure in *b* would result. We would get figure *c* from holding the flashlight above the object. This same tactic can be applied to higher-dimensional figures, producing in each case a set of two-dimensional pictures that represent the whole without loss of information. In each case it is necessary to disregard all but two of the dimensions, which results in a large number of pictures if there are very many dimensions. For example, a 10-dimensional object would require 45 two-dimensional pictures to represent it completely. The saving feature of all this is that there are seldom more than four or five dimensions worthy of notice in a multivariate analysis.

**Geometric terminology.** All the geometry we need for our later discussions is a simple generalization of the principles of the Cartesian coordinate system and Euclidean geometry. We start with an *origin*, which is a point in space. *Coordinate axes* are lines through the origin that are at right angles to each other. A set of these coordinate axes, which are marked off in the units of representation being used (usually standard scores or proportions of variance), form a *coordinate space* or simply a *space*. The number of coordinate axes defines the *dimensionality* of the space. For example, the ordinary scatter plot of the relationship between two variables with the variables represented by the axes is a two-dimensional space or 2-space. If three variables are being represented by three axes through the same origin and all are at right angles to each other, the result is a 3-space. For spaces of higher dimensionality the term *hyperspace* is generally used. In general, the prefix *hyper-* means that something with more than the usual number of dimensions is being discussed. A *hyperplane* is a surface of more than two dimensions; a *hypersphere* is a spherical figure (i.e., one with all points on its surface equidistant from the center) with more than three dimensions. A *hyperellipsoid* is an elliptically shaped figure of four or more dimensions or axes.

Within any given space of two or more dimensions there is an infinite number of other spaces, called *subspaces*, which are of lower dimensionality but which have the same origin. In a plane (a 2-space) there are infinitely many lines (1-spaces) that can be drawn through the origin. In a sphere or 3-space there are infinitely many planes through the origin that are proper subspaces of the original space. There are also infinitely many lines that are proper subspaces of the sphere. All hyperspaces also have an infinite number of subspaces ranging in dimensionality from 1 to $n - 1$, where $n$ is the dimensionality of the hyperspace. The whole field of multivariate analysis can be viewed as the determination of the properties of the best subspace to represent a given hyperspace. What we are doing in all forms of multivariate analysis is trying to account for $n$-dimensional relationships using $n - 1$ or fewer dimensions.

**Geometry of matrices.** We have seen in Chapter 2 that a vector has two interpretations, both of which have geometric implications. In one sense a vector is seen as the list of scores of $N$ individuals on a variable. If we have two such vectors of scores representing the scores of the $N$ people on two variables we could represent each variable as an axis, and by considering a person's two scores simultaneously we could represent him as a point that reflects both his scores. When this is done for all people we get a scatter plot of relationship between the two variables.

Because a matrix, particularly a score matrix, is simply a collection of vectors, it would seem that we should be able to apply the same procedure to a score matrix to obtain a scatter plot of the overall relationship. In fact, we have already done so. Our two vectors of scores form an $N \times 2$ score matrix, which we have represented geometrically in our scatter plot. The same principle can be applied to a score matrix representing the

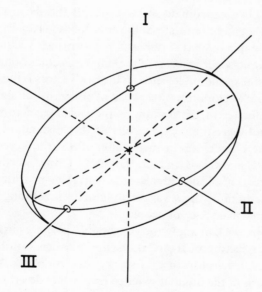

*Figure 4-1   Theoretical 3-dimensional standard score scatter plot. Each point (points are not shown) represents a person. Points are located by the scores of the people on the three variables.*

scores of $N$ people on $n$ variables by using an $n$-dimensional coordinate space in which each dimension represents one variable. We can also show each possible bivariate relationship by simply disregarding all of columns of the matrix except the pair with which we are concerned and by doing this for each possible pair in turn.

Figure 4-1 may help to illustrate these notions. Here we have a three-dimensional scatter plot in which each point represents a person who has been located by his three scores. The figure is the familiar scatter plot ellipse, but in 3-space it resembles a football. If we were to pass a plane through axes I and II, thereby taking a subspace that disregards or is independent of variable III because it is perpendicular to variable III, and project all the points in the ellipsoid onto that plane, we would have the bivariate scatter plot between variables I and II exactly. This is equivalent to using only the first two columns of the score matrix. This has been done in Figure 4-2a. In Figure 4-2b the same procedure has been applied to Figure 4-1 to give the relationship between variables II and III. A similar treatment could provide the I–III scatterplot.

The second interpretation of vectors discussed in Chapter 2 is that a vector is a line in a coordinate space of people that begins at the origin and has its outer end or terminus at the coordinates given by the standard scores of the people on that variable. When two such vectors are plotted in the same people-space in terms of the standard scores obtained by those people, the cosine of the angle between the two vectors is the pmc between them. The score vectors (the first sense of the term) of standard scores give the coordinates of the termini of the variable vectors (the second sense). It would therefore seem reasonable that a standard score *matrix* should give the coordinates of the termini of $n$ variable vectors in an $N$ space of people, and this is, in fact, the case. In other words, a matrix of standard scores for $N$ individuals on $n$ variables provides the coordinates of the termini of the $n$ variable vectors representing the variables in the $N$-dimensional space of people.

At this point we can apply the concept of subspaces in a special way to simplify our picture of the relationships between variables. We have plotted the variable vectors in an $N$ space, but it is not necessary to retain all $N$ dimensions of the original space to completely describe the relationships among the $n$ variables. In fact, we need at most $n$ independent dimensions to retain all the necessary information, because the $n$ variable vectors can always be described by their projections on a set of $n$ arbitrary

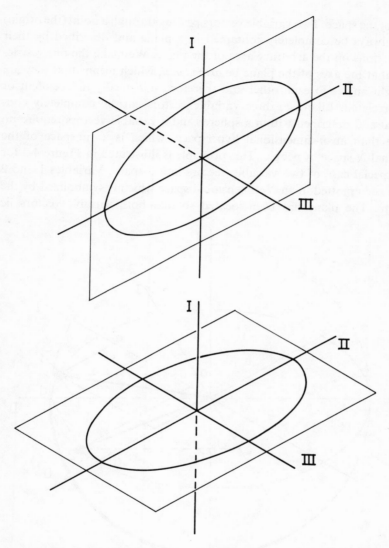

*Figure 4-2   Two projections of the 3-dimensional scatter plot in Figure 4-1 into two dimensions to show bivariate relationships. The bivariate plot of variables I and III is not shown.*

axes of the space. Two variable vectors passing through a point (the origin) can always be completely contained in a plane and described by their projections on the arbitrary axes of the plane. We make the one restriction that the axes of the plane be *orthogonal*, which means that they are at right angles to each other and therefore statistically independent or uncorrelated. Likewise, three variable vectors can be completely contained and described within a sphere; and *n* variable vectors require no more than an *n*-dimensional hyperspace, which is a subspace of the original *N* space of people. This principle is illustrated in Figure 4-3 for the special case of two variable vectors in a 3-space. Variables 1 and 2 have been plotted in an *N*-dimensional space which is symbolized by the sphere. The plane has been located so that both variable vectors lie

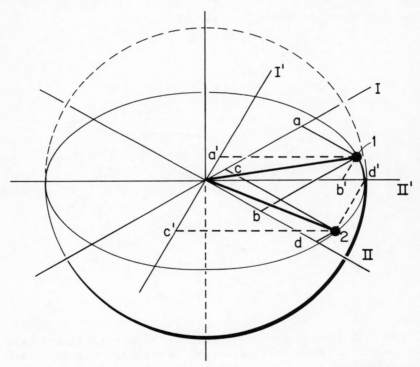

*Figure 4-3    Plot of 2 variables in an N-space. A subspace of two dimensions completely contains the variables. Any pair of axes in the plane (2-space) is equally good for describing the variables.*

completely in it. Axes I and II represent one pair of arbitrary reference axes that can be used to completely describe the two variable vectors. The solid lines and the points $a$ and $b$ give the perpendicular projections of variable vector 1 on axes I and II, and they are also the coordinates of the terminus of vector 1 in the space defined by the axes. The other solid lines and points $c$ and $d$ give the same information for variable vector 2. Axes I′ and II′ are a different pair of reference axes, which also completely describe the plane. Points $a′$, $b′$, $c′$, and $d′$ and their associated dashed lines give the projections of the variable vectors and their coordinates in terms of this different set of reference axes. Either pair of reference axes is equally good for describing the two variable vectors and the relationship between them. Note, however, that it is the location of the reference axes, not the variable vectors, that is arbitrary. The location of the latter is determined by the $N$ space of people used to plot them originally, and it cannot be altered. The subspace must be chosen to include the variable vectors, and it is only the location of the axes within this subspace that is arbitrary.

There is one other analogue of the relationship between the two kinds of vectors with which we must deal before we turn our attention to multivariate procedures. We saw in Chapter 2 that if we cross-multiplied the elements of one standard score vector by the elements of the other, summed the products, and divided by $N$, the resulting value was the correlation between the two variables. It is a simple matter to generalize this procedure to standard score matrices, because the process of matrix multiplication involves cross-multiplication and summation of vectors. If $\mathbf{Z}$ is an $N \times n$ standard score matrix, as shown below

$$\mathbf{Z} = \begin{bmatrix} Z_{11} & Z_{12} & \dots & Z_{1n} \\ Z_{21} & \cdot & \dots & Z_{2n} \\ \vdots & \vdots & & \vdots \\ Z_{N1} & \cdot & \dots & Z_{Nn} \end{bmatrix}$$

then by carrying out the following operations

$$\mathbf{Z}'\mathbf{Z}\left(\frac{1}{N}\right) = \begin{bmatrix} Z_{11} & Z_{12} & \dots & Z_{N1} \\ Z_{12} & & & \\ \vdots & & & \\ Z_{1n} & \cdot & \dots & Z_{Nn} \end{bmatrix} \begin{bmatrix} Z_{11} & Z_{12} & \dots & Z_{1n} \\ \vdots & & & \\ \vdots & & & \\ Z_{1N} & \cdot & \dots & Z_{Nn} \end{bmatrix} \left(\frac{1}{N}\right)$$

we get the correlation matrix

$$\mathbf{R} = \begin{bmatrix} r_{11} & r_{12} & \cdots & r_{1n} \\ r_{21} & r_{22} & \cdots & r_{2n} \\ \vdots & & & \vdots \\ r_{n1} & \cdot & \cdots & r_{nn} \end{bmatrix}$$

where the elements on the main diagonal ($r_{11}, r_{22}, \ldots, r_{nn}$) are the correlations of each variable with itself (1.00), and the off-diagonal elements ($r_{12}, r_{13}, r_{23}$, etc.) are the correlations among the $n$ variables. The correlation matrix is, of course, symmetric, with $r_{12}$ being equal to $r_{21}$, and so on. Notice that the order of multiplication is important. The equation

$$\mathbf{ZZ'}\left(\frac{1}{n}\right) = \mathbf{P}$$

yields an $N \times N$ matrix of the relationships among people, whereas the equation

$$\mathbf{Z'Z}\left(\frac{1}{N}\right) = \mathbf{R}$$

yields the desired $n \times n$ matrix of correlations among variables.

We can see from the foregoing development that the principle we found to hold for pairs of variables and their associated vectors can also be applied to the more complex situation of a standard score matrix. Just as the single correlation coefficient resulting from the multiplication of two score vectors gave us the cosine of the angle between the two variable vectors in a 2-space, the multiplication of two score matrices gives us the $n \times n$ correlation matrix, the elements of which provide us with the cosines of the angles between the $n$ variable vectors in an $n$ space. The correlation matrix contains all the information about the relationships among the $n$ variables.

There are two restrictions we must place on our unbounded enthusiasm for the correlation matrix as an answer to all our problems. First, there is some loss of information. By reducing the score matrix to a correlation matrix we lose information about the $N$ space of people with which we started. It is no longer possible to recover information about the scores of individuals on the variables. However, this loss need not concern us particularly, because we always lose information whenever we summarize a set of scores (e.g., when we use the mean and standard deviation to summarize a distribution). It is a price we are willing to pay for the increased ease with which we can interpret our data. It is much simpler to have single coefficients to describe the relationships between sets of scores than to deal with the unreduced data.

The other restriction we must place on the score matrix is that it have more rows than columns or, more specifically, that there be more people than there are variables. This condition is necessary bècause there must be at least $n + 1$ dimensions in the "people space" for there to be an $n$-dimensional subspace. If there are not at least $n + 1$ people, there will be linear dependence among the variables that will result in the correlation matrix being singular. When this occurs it is not possible to perform the multivariate analyses described later.

Linear dependence exists whenever it is possible to completely describe a set of $n$ variables in a space of $n - 1$ or fewer dimensions. When this is possible, at least one of the variables can be completely described by its relations with the other variables. It contributes no unique information and is therefore redundant with the others. We thus have two reasons for requiring that $N$ be at least as large as $n + 1$; if it is not, there can be no proper subspace of $n$ dimensions, or, if there is a proper subspace of $n - 1$ of fewer dimensions, at least one of the variables will be redundant with the others.

A third reason comes from inferential statistics. When $n$ equals or exceeds $N$ we have what amounts to negative degrees of freedom as that term is used in $t$ tests and the analysis of variance. This, of course, makes any generalization beyond the data at hand impossible. In fact, as we shall see later, unless $N$ is considerably larger than $n$, we are on very unsafe ground in attempting to generalize beyond the sample that provided our data.

From the foregoing discussion the reader may have sensed that problems may arise even when $N$ exceeds $n$. There are some occasions when there will be experimental dependence among the variables. This occurs when one variable is a combination of others, such as including height, weight, and the ratio of height to weight in the same analysis. Such practices should be avoided in multivariate analysis, because they will cause the correlation matrix to be singular or nearly so, thus making the determination of the inverse impossible or seriously affecting its accuracy. The latter is the more dangerous case, because its effects may go undetected.

A second way in which dependence may be introduced into the data is in the method used for collecting responses. We refer here to the forced-choice format in its various manifestations. The most obvious case is that of a complete paired-comparisons data collection procedure in which the subjects are asked to choose between *A* and *B*, between *A* and *C*, between *B* and *C*, and so forth. The result is a set of scores that are *ipsative* (in that they provide a complete ranking of the variables for each individual) and are therefore linearly dependent. That is, the scores for all individuals sum to a constant, thus any one score for an individual is that constant minus the sum of his other scores. Related methods of data collection such as the method of multiple rank orders can also yield ipsative data. The characteristics of ipsative data are not well understood, but it is known that they may cause problems of partial or complete linear dependence among variables. It is also known that circular triad analysis of comparative judgment data can yield psychometric dividends. The potential benefits and hazards of ipsative data should be considered before data collection to determine whether the psychometric benefits are worth the multivariate price. (See Coombs, 1964 for a discussion of ranking methods of data collection.)

We have seen one special matrix operation in which the score matrix **Z** is premultiplied by its transpose and postmultiplied by $1/N$ to yield the correlation matrix **R**, and we have seen the geometric consequences of this. It is useful at this point to examine what happens geometrically when other matrix manipulations are performed. We will take two simple $2 \times 3$ matrices for illustration, because they will enable us to depict the geometric results in visual form.

The first readily apparent principle is that it makes a difference which way we read the matrix. We have already seen this principle in action with our score matrix. If we consider people (rows) as points and variables (columns) as axes, we get a set of $N$ points in a space of $n$ axes. Looking at variables as points and people as axes, we obtain an $N$ space containing $n$ points. The same thing will hold for the $2 \times 3$ matrix, **A**, below. Figure 4-4a shows the result of considering rows to be dimensions (call this the matrix **A**); the alternative case is shown in Figure 4-4b (call this matrix **A**′).

$$\mathbf{A} = \begin{bmatrix} 1 & -1 & 2 \\ -2 & -2 & 2 \end{bmatrix}$$

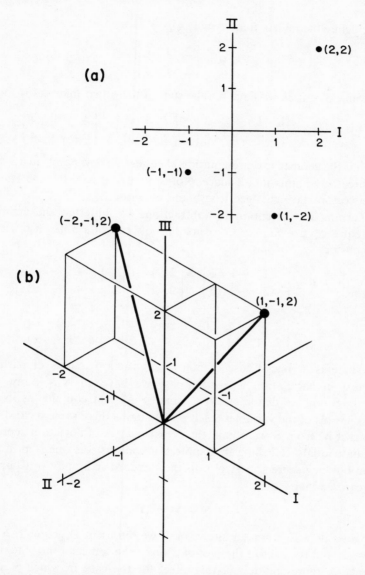

*Figure 4-4   Two geometric representations of the same score matrix. In part (a) people are plotted as points in a space of variables. In part (b) variables are plotted as points (and vectors) in a space of people.*

If we define the matrix **B** as

$$\mathbf{B} = \begin{bmatrix} 2 & 1 & 1 \\ 2 & 1 & 2 \end{bmatrix}$$

we could plot it as we have **A**. The sum of these two matrices is

$$\mathbf{A} + \mathbf{B} = \begin{bmatrix} 1 & -1 & 2 \\ -2 & -2 & 2 \end{bmatrix} + \begin{bmatrix} 2 & 1 & 1 \\ 2 & 1 & 2 \end{bmatrix} = \begin{bmatrix} 3 & 0 & 3 \\ 0 & -1 & 4 \end{bmatrix}$$

and has the geometric representation of either two new points in a 3-space or three new points in a 2-space. Notice that $\mathbf{A} + \mathbf{B} = \mathbf{B} + \mathbf{A}$. The variable vectors have changed length and direction.

Turning our attention to subtraction, we find that the geometric consequences are similar. Of course, it now makes a difference whether we consider

$$A - B = \begin{bmatrix} 1 & -1 & 2 \\ -2 & -2 & 2 \end{bmatrix} - \begin{bmatrix} 2 & 1 & 1 \\ 2 & 1 & 2 \end{bmatrix} = \begin{bmatrix} -1 & -2 & 1 \\ -4 & -2 & 0 \end{bmatrix},$$

or

$$B - A = \begin{bmatrix} 2 & 1 & 1 \\ 2 & 1 & 2 \end{bmatrix} - \begin{bmatrix} 1 & -1 & 2 \\ -2 & -2 & 2 \end{bmatrix} = \begin{bmatrix} 1 & 2 & -1 \\ 4 & 3 & 0 \end{bmatrix}$$

There is a special case of matrix subtraction that is of particular interest in correlation work. Suppose we have an $N \times n$ raw score matrix **S** and we define an $N \times n$ matrix **M** that contains as its rows the $n$ means of the variables (each row contains these same means). If we subtract **M** from **S** we obtain the $N \times n$ matrix of deviation scores for the individuals. Recalling that subtracting means was one step in going from our raw score scatter plot to the standard score plot in Chapter 2, we can see that

$$\mathbf{S} - \mathbf{M} = \mathbf{D}$$

provides us with a similar situation. If we represent **D** geometrically we have moved the axes of the plot of **S** up to the center of the scatter of $N$ points. A representation of this effect for the case in which **S** and **D** are $N \times 3$ matrices is shown in Figure 4-5.

Multiplication of matrices can have more complicated geometric consequences than addition or subtraction because unless one of the matrices is square, the dimensionality of the space will be changed.

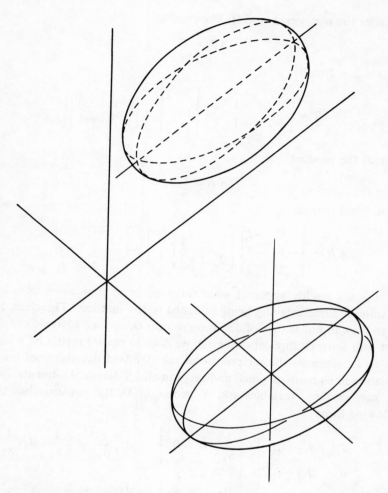

*Figure 4-5   What happens to the scatter plot when a raw score matrix is converted into a deviation score matrix. The upper ellipsoid is the raw score plot.*

Consider the matrices **A** and **B**. The product

$$AB' = C$$

yields a 2 × 2 matrix

$$AB' = \begin{bmatrix} 1 & -1 & 2 \\ -2 & -1 & 2 \end{bmatrix} \begin{bmatrix} 2 & 2 \\ 1 & 1 \\ 1 & 2 \end{bmatrix} = \begin{bmatrix} 3 & 5 \\ -3 & -1 \end{bmatrix}$$

whereas the product

$$A'B = C^*$$

is the 3 × 3 matrix,

$$A'B = \begin{bmatrix} 1 & -2 \\ -1 & -1 \\ 2 & 2 \end{bmatrix} \begin{bmatrix} 2 & 1 & 1 \\ 2 & 1 & 2 \end{bmatrix} = \begin{bmatrix} -2 & -1 & -3 \\ -4 & -2 & -3 \\ 8 & 4 & 6 \end{bmatrix}$$

It is not readily apparent what this type of multiplication does for us without some meaning being attached to the matrices. Therefore, let us go back to our matrix of deviation scores, **D**, and see what we can do to it with some multiplications. First, we define a square matrix **S\*** which has as its elements the reciprocals of the standard deviations of the $n$ variables in its main diagonal and zeros in all off-diagonal elements. For the case of three variables and $N$ people in **D**, the equation has the following form

$$DS^* = \begin{bmatrix} d_{11} & d_{12} & d_{13} \\ d_{21} & d_{22} & d_{23} \\ \vdots & & \\ d_{i1} & \cdots & d_{i3} \\ \vdots & & \\ d_{N1} & \cdots & d_{N3} \end{bmatrix} \begin{bmatrix} \dfrac{1}{S_1} & 0 & 0 \\ 0 & \dfrac{1}{S_2} & 0 \\ 0 & 0 & \dfrac{1}{S_3} \end{bmatrix} = \begin{bmatrix} d_{11}\dfrac{1}{S_1} & d_{12}\dfrac{1}{S_2} & d_{13}\dfrac{1}{S_3} \\ d_{21}\dfrac{1}{S_1} & \cdots\cdots\cdots\cdots \\ d_{i1}\dfrac{1}{S_1} & \cdots\cdots\cdots\cdots \\ d_{N1}\dfrac{1}{S_1} & \cdots\cdots\cdots & d_{N3}\dfrac{1}{S_3} \end{bmatrix}$$

But the elements of the product matrix are deviation scores divided by their standard deviations, which are standard scores. Therefore, we may write

$$(\mathbf{S} - \mathbf{M})\mathbf{S}^*_{\text{diag}} = \mathbf{Z}$$

as a compact expression for finding the standard scores of $N$ individuals on $n$ variables. The parentheses indicate that the subtraction should take place before the multiplication, and $\mathbf{S}^*_{\text{diag}}$ expresses the fact that $\mathbf{S}^*$ is a matrix with nonzero values only in the diagonal. Geometrically, we know from Chapter 2 that the change from deviation scores to standard scores changes the shape of our ellipsoid by equating the amount of spread of scores on each of the three axes.

We have already examined the geometric consequences of another particular matrix multiplication, $\mathbf{Z}'\mathbf{Z}(1/N) = \mathbf{R}$. In this case we considered people to be axes and found that the equation gave us the degree of relationship among the variables (the rows of $\mathbf{Z}'$). We also noted that $\mathbf{Z}\mathbf{Z}'(1/n)$ gave us a different matrix, an $N \times N$ matrix of the relationships among people as defined by the $n$ variables. In general, we may say that a matrix multiplied (either pre or post) by its transpose results in a symmetric matrix indicating the degree of relationship among the entities that form the rows of the first matrix in the equation. This is true regardless of whether the elements of the matrix are raw, deviation, or standard scores. $\mathbf{S}'\mathbf{S}$ yields a matrix of crossproducts among the raw scores of the general form

$$\sum_{i=1}^{N} X_i Y_i$$

$\mathbf{D}'\mathbf{D}$ gives us covariancelike terms of the form

$$\sum_{i=1}^{N} x_i y_i$$

and, of course, $\mathbf{Z}'\mathbf{Z}$ gives us values proportional (by a constant of $1/N$) to the correlations. In general, we will find it most convenient to work with this last form, because the scale of relationships among the points is not affected by differences in the variances of the variables. Note that $\mathbf{S}\mathbf{S}'$, $\mathbf{D}\mathbf{D}'$, and $\mathbf{Z}\mathbf{Z}'$ provide the same types of information about the relationships between people.

Finally, we examine the geometric meaning of the inverse of a matrix. The various relationship matrices described above ($S'S$, $D'D$, and $Z'Z$) have sums of squares in their diagonals. Considering the latter two, we obtain

$$\sum_{i=1}^{N} x_{ij}^2 \quad \text{and} \quad \sum_{i=1}^{N} z_{ij}^2$$

as the $j$th diagonal elements, respectively. When these values are multiplied by $1/N$, we obtain the variance of variable $j$ in raw and standard score form, respectively, in the diagonal. (We can do the same for $S'S$ if we make an adjustment for the means of the variables.) The off-diagonal elements are indices of relationship expressed as covariances and correlations. Either of these matrices can be used to plot the $n$ variable vectors in an $n$ space, and these vectors will have length equal to the square root of their respective diagonal elements.

Consider what the plot of an indentity matrix would look like. Because the identity matrix contains 1.0 in its diagonal (for vector length) and zeros everywhere else (for vector relationships), it is apparent that the plot of $I$ will be a set of $n$ vectors, each of length 1.0 and each at right angles to all other vectors. Thus the identity matrix defines the $n$ independent axes of an ordinary $n$-dimensional Cartesian space. Because the inverse of a matrix is defined as another matrix that, when multiplied by the first matrix, yields $I$, the plot of the inverse must be a set of vectors that transforms the original vectors to unit length and independence of each other. Geometrically, the inverse of a matrix will generally result in a set of $n$ vectors radiating from the origin in more or less the opposite direction from the original set of vectors.

## SUMMARY

We have seen in this chapter that a matrix is nothing more nor less than a simple way of keeping track of a group of numbers that can be categorized on two or more dimensions. A matrix has the properties of order, which is the size of the matrix, and rank, which refers to the number of unique sources of information in the matrix.

In multivariate statistics we deal with two types of primary matrices, score matrices and the correlation matrices derived from them, and various types of secondary matrices. All these matrices have geometric meanings that are used to develop an understanding of the interpretations of the matrices themselves. A special type of score matrix, the standard score matrix, can be used both to prepare a multidimensional scatter plot of the usual form, in which the axes represent variables and the points are people, and to plot the variable vectors in a space with the axes representing people. In this latter case the cosines of the angles between the variable vectors give the correlations among the variables. The correlation matrix gives these cosines directly. There is an $n$-dimensional subspace of the $N$ space of people that can completely describe the relations among the variables. We require that there be more people than variables to avoid the problem of linear dependence.

# PART II
## EXTERNAL FACTOR ANALYSIS

# 5

*Semi-partial*

## PART AND PARTIAL CORRELATION

In Chapters 2 and 3 we discussed a variety of coefficients that could be used to describe the relationship between a pair of variables or to predict the scores of individuals on one variable from their scores on another. There are many occasions, however, when more than two variables are available for study. In this section on external factor analysis we discuss methods for use when prediction *may* be a goal and more than two variables are available.

### PART CORRELATION

Suppose that we have measured a group of people (hopefully, it is a random sample from a defined population, but in practice it probably is not) on three variables we feel will help to unravel the mysteries of academic performance. These variables are called academic achievement (variable *A*), academic potential (variable *P*), and academic interest (variable *I*). If we have a theory about the nature of academic achievement which says that achievement in academic settings is a function of one's academic potential and one's interest in academic subjects and activities, these variables would be important to our theory. By the technique of multiple correlation we could determine the degree to which achievement is predictable from both potential and interest. However, we may be interested also in certain bivariate relationships between these variables. For example, we might wish to know the relationship between achievement and potential when variation in achievement which is related to variation in interest is held constant. That is, we might wish to determine the degree of relationship between achievement and potential when any association between achievement and interest is removed. Alternatively, we might wish to remove variation in achievement which is related to potential from the relationship between achievement and interest. Either of these questions can be answered by a technique known as *part correlation*. Another name for the procedure we are about to discuss is *semipartial correlation*. The reason for this name will become apparent shortly.

There are two approaches to help us understand what is achieved by part correlation. First, let us consider the situation where we wish to determine the relationship between achievement and potential with the effect of interest on achievement removed. We can form a scatter plot of the interest and achievement scores, as shown in Figure 5-1. By forming

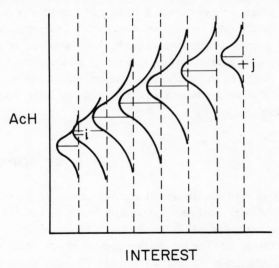

*Figure 5-1   Distributions of achievement scores after grouping on interest.*

groups of subjects based on their interest scores, which is illustrated by the small normal distributions in the figure, and by representing each person's achievements as a deviation from the mean achievement of his interest level group, we have succeeded in developing an achievement score that is free of the influences of interest. That is, each person's $A$ score is reported relative to the $A$ scores of others with similiar interest levels.

To see the effect of changing from raw achievement scores (these raw achievement scores might be in standard score form, but we refer to them as raw scores to reflect the fact that they will be modified subsequently) to scores that are deviations from subgroup means, consider the scores of individuals $i$ and $j$. In terms of raw scores, $j$ has a higher score than $i$; however, when we look at the relative position of each person within his group we find that $j$'s achievement score is substantially below the mean of achievement scores of people with interest scores similar to his, whereas $i$ is considerably above his group's mean. Thus, with variation in interest removed (or level of interest held constant), $i$ has a higher achievement score than $j$. To find the part correlation, then, between achievement and potential with the effects of variation in interest on achievement removed, we find the product moment correlation between *raw scores on potential and deviation scores on achievement*, where the deviation scores are deviation from achievement means for interest level subgroups.

The procedure described above would be tedious and rather inaccurate to use in practice; tedious because it would be necessary to compute all those means and deviation scores, and inaccurate because of the

**TABLE 5-1**

*Computation of Part Correlations with Interest Held Constant.*

|   | $A$ | $P$ | $I$ |
|---|------|------|------|
| $A$ | 1.00 | .60 | .40 |
| $P$ | .60 | 1.00 | .30 |
| $I$ | .40 | .30 | 1.00 |

$$r_{P(A \cdot I)} = \frac{.60 - .30(.40)}{\sqrt{1 - (.40)^2}} = \frac{.48}{\sqrt{.84}} = .52$$

$$r_{A(P \cdot I)} = \frac{.60 - .40(.30)}{\sqrt{1 - (.30)^2}} = \frac{.48}{\sqrt{.91}} = .50$$

errors involved in grouping. The inaccuracy could be removed by using deviations from the regression line for predicting achievement from interest in place of deviations from calculated group means, but either procedure is unnecessary, because there is a simple way to compute the part correlations directly from the original correlations. Equation (5.1) is the formula to use for finding the part correlation between $A$ and $P$ with the effect of $I$ removed from $A$. In this equation, $r_{P(A \cdot I)}$ is the correlation between $A$ and $P$ with $I$ held constant for $A$.

$$r_{P(A \cdot I)} = \frac{r_{PA} - r_{PI}r_{AI}}{\sqrt{1 - r_{AI}{}^2}} \qquad Semi\text{-}partial \qquad (5.1)$$

The terms $r_{PA}$, $r_{PI}$, and $r_{AI}$ are the regular pmc's between $P$ and $A$, $P$ and $I$, and $A$ and $I$, respectively. The results from using this formula are exactly the same as those that would be obtained if $A$-score deviations from the regression line for predicting $A$ from $I$ were correlated with $P$ scores.

The reader should note that there is another part correlation between $P$ and $A$ that could be computed from the same data, namely, the correlation between potential and achievement with interest held constant in potential. This part correlation would be given by the formula

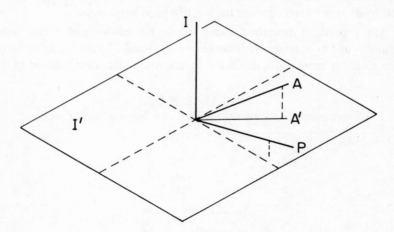

*Figure 5-2   Vector representation of the part correlation between potential and achievement with interest "parted out" of achievement. The correlation is between P and A'.*

$$r_{A(P \cdot I)} = \frac{r_{AP} - r_{AI}r_{PI}}{\sqrt{1 - r_{PI}^2}}$$

The only substantive difference between this formula and equation (5.1) is the correlation squared in the denominator. Table 5-1 shows the working out of the part correlations for the two examples discussed, using hypothetical data. Note that there are four other possible part correlations that could be computed, the most interesting of which would probably be $r_{I(A \cdot P)}$.

There is another way to view what is going on in part correlation which we will find quite useful when our discussion turns to more complex statistical procedures. This alternative approach develops out of the geometric interpretation of correlation. If we were to represent the relationships found in Table 5-1 geometrically by variable vectors in a space of people, the vectors would look like those in Figure 5-2. Because we are dealing with standard scores and the length of each vector is its standard deviation, each vector is of unit length and the angles between the vectors reflect the correlations between them. Because the variables are all positively correlated, their vectors form acute angles. There is one and only one plane in the 3-space that passes through the origin and *is perpendicular to the vector representing variable I*. This plane has been drawn in as plane $I'$. Because vector $I$ is perpendicular to plane $I'$, the plane is independent of the vector (orthogonal to it). If we take our flashlight again (as described in Chapter 4) and project vector $A$ down onto plane $I'$, represented in the figure by the line $A'$, then $A'$ represents that portion of $A$ which is independent of $I$. $A'$ will not be of unit variance because that variance it shared with $I$ has been removed. Therefore, the scores people have on $A'$ are no longer standard scores but are scores that have a variance less than unity. (The dashed line below $P$ is used to indicate that $P$ is above the plane.)

The reader will recall from Chapter 2 that the cosine of the angle between two variable vectors is equal to the correlation between the two variables only when the variables have unit variance, yielding vectors with lengths equal to one. When this is not the case, the proper part correlation must include a correction for reduced length. This correction component is found in the denominator of equation (5.1). The reader should recognize this correction term as the standard error of estimate of standardized $A$ scores when predicted from $I$ scores, which is, of course, the standard deviation of $A'$ scores or the length of the vector $A'$. Thus the part correlation can still be viewed as the cosine of the angle between two vectors; however, now one of the vectors has changed so that it has less than unit variance.

There is an interesting parallel between what we accomplish by computing a part correlation and certain sampling procedures used in inferential statistics. In a sense, computing a part correlation is very similar to the use of control variables in an experimental design. That is, by using a part correlation the effect of some third variable is removed from the relationship between two others. By the same token, holding a variable constant in sampling, such as using psychiatric patients who all have the same type and severity of symptoms, removes the possibility of variation on this dimension influencing the results of a study. The situation is even more nearly parallel in the case of partial correlation, to which we now turn.

## PARTIAL CORRELATION

Suppose that we continue to hold our theory about achievement being a result of potential and interest. However, let us now hypothesize that potential is a precursor to both achievement and interest. It might be of interest for us to determine what the correlation is between achievement and interest, with the effect of variation in potential removed from both these variables. The solution to this problem is known as partial correlation, and its logic is essentially the same as that of part correlation. (At this point the reason for semi-partial as an alternate term for part correlation should be obvious, because it deals with half of a partial correlation in a logical sense.)

In computing a partial correlation between $A$ and $I$ with $P$ partialled out of the relationship what we are doing, in essence, is forming groups according to scores on $P$, finding the mean of $A$ scores and the mean of $I$ scores for the members of each group, and assigning to each person as his $A'$ score the deviation of his $A$ score from the group's $A$ mean and as his $I'$ score the deviation of his $I$ score from the group's $I$ mean. The partial correlation between $A$ and $I$ is the correlation between the pairs of deviation scores, $A'$ and $I'$. More precisely, each person's $A'$ is the deviation of his $A$ score from the regression line for predicting $A$ from $P$, and his $I'$ is the deviation of his $I$ score from the regression line for predicting $I$ from $P$. The partial correlation $r_{AI \cdot P}$ is the correlation between these two sets of deviations. The subscript $AI \cdot P$ is used to indicate that the effect of $P$ is being removed from (or held constant in) both $A$ and $I$.

The formula for computing the partial correlation is

$$r_{AI \cdot P} = \frac{r_{AI} - r_{AP} r_{IP}}{\sqrt{1 - r_{AP}^2} \sqrt{1 - r_{IP}^2}} \qquad (5.2)$$

in which each of the terms has the same meaning it had for computing the part correlations. The value for this partial correlation, using the data from Table 5-1, is

$$r_{AI \cdot P} = \frac{.40 - .60(.30)}{\sqrt{1 - .60^2} \sqrt{1 - .30^2}} = \frac{.40 - .18}{.80(.954)} = \frac{.22}{.763} = .29$$

To put the partial correlation into perspective with regard to the part correlation, the partial correlation, $r_{AP \cdot I}$ (the same relationship computed in Table 5-1), is

$$r_{AI \cdot P} = \frac{.60 - .40(.30)}{\sqrt{1 - .40^2} \sqrt{1 - .30^2}} = \frac{.48}{.917(.954)} = \frac{.48}{.875} = .55$$

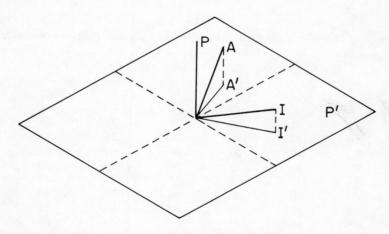

*Figure 5-3   Vector representation of the partial correlation between interest and achievement with potential "partialled out." The correlation is between $A'$ and $I'$.*

The partial correlation is also similar to the part correlation geometrically. The only difference is that both variable vectors are projected onto the plane that is orthogonal to the vector of the variable being partialled out. Figure 5-3 depicts the geometric situation when $P$ is being partialled out of the relationship between $A$ and $I$. Here, the $P'$ plane is orthogonal to the $P$ vector. $A$ and $I$ lie above the plane, whereas $A'$ and $I'$ are the projections of $A$ and $I$ onto plane $P'$. Both $A'$ and $I'$ now have less than unit length, and the correction for their reduced variance is incorporated as the two correction components, both standard errors of estimate, in the denominator of equation (5.2).

Some special cases of part and partial correlations may help in developing an understanding of how these coefficients behave. In Table 5-2 we have computed the one partial correlation and two part correlations that are possible when the effect of variable 3 is removed from either or both of two other variables.

One of the first and perhaps most surprising things we find in Table 5-2 is that a part or partial correlation may be nonzero, even when the whole correlation is zero. We also see that when the sign of either $r_{13}$ or $r_{23}$ (but not both) is negative, the part and partial correlations are larger than the whole correlation, but when both $r_{13}$ and $r_{23}$ have the same sign, the part and partial correlations are smaller. When $r_{13}$ is zero, $r_{2(1.3)} = r_{12}$ because variable 1 is already in the plane perpendicular to variable 3.

### TABLE 5-2
*Effect of Different Raw Correlations on the Part and Partial Correlations.*

| Raw Correlations | | | Part Correlations | | Partial |
|---|---|---|---|---|---|
| $r_{12}$ | $r_{13}$ | $r_{23}$ | $r_{1(2.3)}$ | $r_{2(1 \cdot 3)}$ | $r_{12 \cdot 3}$ |
| .00 | .50 | .50 | −.29 | −.29 | −.33 |
| .00 | −.50 | .50 | .29 | .29 | .33 |
| .20 | .50 | .50 | −.06 | −.06 | −.07 |
| .40 | .50 | .30 | .26 | .29 | .30 |
| .60 | .30 | .60 | .52 | .44 | .55 |
| .60 | −.30 | .50 | .86 | .79 | .90 |
| .80 | .00 | .50 | .92 | .80 | .92 |
| .80 | .60 | .60 | .55 | .55 | .69 |
| .80 | −.60 | −.60 | .55 | .55 | .69 |

Finally, we notice that the partial correlation, disregarding sign, is always at least as large as either part correlation, a fact that arises because both variables are being projected into the same plane in partial correlation.

## TESTS OF SIGNIFICANCE

Testing the significance of a partial correlation (i.e., estimating the probability that the obtained partial correlation is a chance deviation from 0) can be done by the Fisher $Z$ transformation in which the standard error of the $Z$ corresponding to the given partial $r$ is $1/\sqrt{N-4}$. Thus if we assume that the data in Table 5-1 were obtained from 40 people, the standard error of the partial correlations $r_{AI \cdot P}$ and $r_{AP \cdot I}$ would be

$$\frac{1}{\sqrt{N-4}} = \frac{1}{\sqrt{36}} = \frac{1}{6} = .167$$

Using the $Z$ transformation with $r_{AI \cdot P} = .29$ and $r_{AP \cdot I} = .55$,

$$Z_{r_{AI \cdot P}} = .299 \qquad Z = \frac{.299}{.167} = 1.79$$

and

$$Z_{r_{AP \cdot I}} = .618 \qquad Z = \frac{.618}{.167} = 3.7$$

From these tests we would probably conclude that $r_{AI \cdot P}$ might be a chance deviation from a population value of zero because it fails to reach the .05 two-tail level of significance. On the other hand, $r_{AP \cdot I}$ is almost certainly nonzero.

An alternative test of significance for use with small samples is given by McNemar (1969). Using the formula

$$t = \frac{r_{12 \cdot 3}}{\sqrt{(1 - r_{12 \cdot 3}^2)/(N-3)}}$$

we would obtain for the example above

$$t_{r_{AI \cdot P}} = \frac{.29}{\sqrt{\dfrac{.916}{37}}} = \frac{.29}{\sqrt{.025}} = \frac{.29}{.16} = 1.81$$

and

$$t_{r_{AP.I}} = \frac{.55}{\sqrt{\dfrac{.7}{37}}} = \frac{.55}{\sqrt{.019}} = \frac{.55}{.14} = 3.9$$

These results agree quite closely with those obtained using the $Z$ transformation formula, as they should. However, the critical values for $t$ should be found from a $t$ table with $N - 3$ degrees of freedom.

The test for the significance of a part correlation requires quite a different formula. McNemar (1969) shows that the proper test for the part correlation is

$$F = \frac{r_{1(2\cdot3)}^2}{(1 - r_{1\cdot23}^2)/(N - 3)}$$

in which $r_{1(2\cdot3)}$ is the part correlation and $r_{1\cdot23}$ is the multiple correlation for predicting variable 1 from a combination of variables 2 and 3. This $F$ has 1 and $N - 3$ degrees of freedom.

Because the significance test for the part correlation is closely related to multiple correlation and requires topics we develop in the next chapter, we refrain from further consideration of this formula until the necessary groundwork has been laid. Part correlations will resurface in connection with our discussion of multiple correlation beta weights.

## HIGHER-ORDER PARTIALS

Sometimes situations arise when it would be desirable to know the correlation between two variables with two or more other variables "partialled out" of the relationship. Let us expand our achievement, potential, interest example to see what this means and how it might be accomplished.

Suppose we add the variable general intelligence ($G$) to our example so that we have data on all four variables. It might be of interest to us to determine the relationship between achievement and potential (assuming that academic potential is something other than general intelligence), with the effects of variation in interest and general intelligence removed from both achievement and potential. This is equivalent to forming categories simultaneously on $I$ and $G$ so that each individual is placed in a single *IG* class, computing the *A* and *P* means within each *IG* level and using deviation scores from these *A* means and *P* means as the scores for computing the correlations. In the regression sense, this is equivalent to using the differences between actual *A* and *P* scores and those scores predicted by multiple regression using *I* and *G* as predictors.

Viewing the situation geometrically requires a four-dimensional space. If we take the plane in the 4-space that is orthogonal to both the *I* and *G* variable vectors, we can project the vectors for *A* and *P* onto this plane. Both the *A* and *P* vectors will have less than unit length, and the cosine of the angle between them corrected for reduced length will be the partial correlation.

The formula for a second-order partial correlation such as the one described above is given by the equation

$$r_{AP \cdot IG} = \frac{r_{AP \cdot I} - r_{AG \cdot I} r_{PG \cdot I}}{\sqrt{1 - r_{AG \cdot I}^2} \sqrt{1 - r_{PG \cdot I}^2}} \tag{5.3}$$

in which the various terms on the right-hand side of the equation are themselves first-order partials. Table 5-3 gives the computational details for the hypothetical data of Table 5-1, with the addition of a row and

**TABLE 5-3**
*Computation of a Second-Order Partial Correlation.*

|   | *A* | *P* | *I* | *G* |
|---|-----|-----|-----|-----|
| *A* | 1.00 | .60 | .40 | .50 |
| *P* | .60 | 1.00 | .30 | .80 |
| *I* | .40 | .30 | 1.00 | .40 |
| *G* | .50 | .80 | .40 | 1.00 |

$$r_{AP \cdot I} = .55 \qquad r_{AG \cdot I} = .40 \qquad r_{PG \cdot I} = .78$$

$$r_{AP \cdot IG} = \frac{.55 - .40(.78)}{\sqrt{1 - .40^2} \sqrt{1 - .78^2}} = \frac{.55 - .31}{\sqrt{.84} \sqrt{.39}} = \frac{.24}{.92(.62)} = \frac{.24}{.57} = .435$$

$$Z_{r_{AP \cdot IG}} = .465 \qquad Z = \frac{.465}{1/\sqrt{35}} = \frac{.465}{.17} = 2.74$$

$$t = \frac{.435}{\sqrt{\dfrac{1 - .435^2}{36}}} = \frac{.435}{\sqrt{\dfrac{.81}{36}}} = \frac{.435}{\sqrt{.0225}} = \frac{.435}{.15} = 2.90 \qquad (\text{df} = 36)$$

column for the new variable, general intelligence. If we had another variable and wished to compute a third-order partial, the formula would be like equation (5.3), but the terms to the right of the equal sign would be second-order partials. Obviously, the computation of higher-order partial correlations quickly becomes a laborious task. The data and common sense seldom warrant partials beyond the second order.

Higher-order partial correlations can be tested for significance by a slight adjustment of the formulas given above for testing first-order partials. For large samples the standard error of $Z$ is given by $1/\sqrt{N-(m+3)}$, where $m$ is the number of variables being partialled out. Likewise, the small sample test becomes

$$t = \frac{r_{12 \cdot 34 \cdots}}{\sqrt{(1 - r_{12 \cdot 34 \cdots}^2)/[N - (m + 2)]}}$$

where $r_{12 \cdot 34 \cdots}$ is the obtained correlation with $m$ variables partialled out of the relationship. Both these tests are performed in Table 5-3 for a hypothetical sample size of 40.

Higher-order part correlations are closely related to a topic in multiple correlation. Because the computational procedures are quite complex, we reserve discussion of this topic until we have developed some of the ideas of multiple correlation.

The final topic of this chapter concerns the possible limits of a correlation between two variables when the correlation of each with a third variable is known. McNemar (1969, p. 185) states that the limits for the situation where $r_{12}$ and $r_{13}$ are known are given by

$$r_{12}r_{13} \pm \sqrt{1 - r_{12}^2 - r_{13}^2 + r_{12}r_{13}}$$

That is, $r_{23}$ will have $r_{12}r_{13} + \sqrt{1 - r_{12}^2 - r_{13}^2 + r_{12}r_{13}}$ as one limit (upper in the case in which all correlations are of the same sign) and $r_{12}r_{13} - \sqrt{1 - r_{12}^2 - r_{13}^2 + r_{12}r_{13}}$ as the other limit. These limits tell us that when $r_{12} = r_{13} = 1.0$, then $r_{23}$ must be 1.0, and when $r_{12} = r_{23} = 0$, then $r_{23}$ can have any value between $+1.0$ and $-1.0$. Less extreme values of $r_{12}$ and $r_{13}$ give intermediate results.

## SUMMARY

In this chapter we have discussed indices of bivariate relationship that can be used when it is desired to remove the effect of a third variable from that relationship. Part or semi-partial correlation refers to the index that results when the third variable is held constant for one of the two variables being investigated, and partial correlations yield the degree of relationship when the effect of the third variable is removed from both of the primary variables of interest.

Both part and partial correlations involve deviations from group means or regression lines involving the third variable. Both are pmc's between appropriately modified sets of scores. A geometric view of these indices was also developed and higher-order partial correlations were described. Tests of statistical significance are available.

# 6

# MULTIPLE CORRELATION

In the preceeding chapter we discussed ways to examine the relationship between two variables while holding the effect of one or more other variables constant. This chapter deals with the other side of the problem and presents methods for examining the relationship between one variable and a combination of two or more other variables that are considered simultaneously. The topic generally goes under the name of *multiple correlation* or *multiple regression.* Some of the techniques described below have been widely used in both research and practice in the behavioral sciences, whereas others, relating principally to external factor analysis, have seldom, if ever, been used. We first consider the simple case of using two variables to predict a third and then go on to more complex examples.

## THREE-VARIABLE CASE

Let us take our example from the preceeding chapter and change our objective from one in which we wish to know the degree of association

139

between ability and potential with the effect of interest removed to one in which we wish to obtain the best possible prediction of *A*, given information about *P* and *I*. This is a problem that confronts high school counselors and college admissions officers daily, and it is one that calls for multiple regression for its solution. Two tasks confront us—what is the best combination of *P* and *I* for predicting *A*, and how good is the prediction? We can attack the two problems separately.

**Combining variables.**    The prediction problem we have posed above calls for the combining of two variables in such a way that the scores resulting from the combination have maximum relationship with the variable to be predicted. The variables that are combined to do the predicting are often called predictors, and the variable to be predicted is called the criterion. Despite the restrictions these terms imply( we need not want to predict anything), they will be used rather than the alternatives of independent and dependent variables, because these latter terms connote the presence of experimental controls and manipulations that seldom exist in practice. We will add the restriction that this best combination of predictors be linear in form, which means that we will not use squared or other powered terms of the predictors in our solution.

First, consider a space defined by the people in our sample in which the variable vectors for *A*, *P*, and *I* have been plotted. The subspace containing these vectors, the same one we had in Figure 5-1, is shown in Figure 6-1, except in this case the space is drawn to represent *P* and *I*

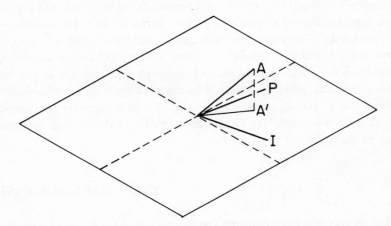

*Figure 6-1    Vector representation of a multiple correlation. A′ is the combination of I and P which has the highest correlation with A.*

as lying in the plane. Although $P$ and $I$ are correlated, they can still be used indirectly to represent any line that lies in the plane. In other words, any line in the plane can be expressed as some simple, linear function of $P$ and $I$. Any line that does not lie precisely in the plane cannot be completely defined by $P$ and $I$. Therefore, the combination of $P$ and $I$ we seek, the one that will predict $A$ with the greatest accuracy (or, more correctly, the smallest error, because our solution will be a least-squares one) must lie in the plane that contains and is defined by $P$ and $I$. We know that when variables are represented as vectors the angle between them is related to their correlation; the smaller the angle, the larger the correlation. The vector in the plane $PI$ that has the smallest angle with the criterion variable $A$ is the line formed when the $A$ vector is projected perpendicularly down onto the $PI$ plane. This line, labeled $A'$ in Figure 6-1, is the combination of $P$ and $I$ that will yield the best possible prediction of $A$, or has the highest relationship with $A$.

The problem that now arises is how to express $A'$ as some linear combination of our two predictor variables. If $P$ and $I$ are not correlated, this presents no problem. We shall see later that the solution for this special case is that each predictor variable receives a weight equal to its correlation with the criterion variable. However, when the predictor variables are correlated, regardless of whether their correlation is positive or negative, we must use only that portion of each predictor variable's variance that is independent of the other predictor variable. We must do this because the representation of a line in a Cartesian space must be in terms of independent reference axes.

We will use the term *predictor composite* to refer to vectors such as $A'$, because they are combinations of several predictor variables. In general, the term *composite* means a latent vector that is a combination of observed variable vectors. A variable vector is one that can be observed directly in the sense that we have a measuring instrument (test, galvanometer, clock, or whatever) from which we get the scores of our subjects directly. A latent vector or composite does not have a measuring instrument associated with it. Rather, the composite is measured indirectly as a combination of scores from several directly measured variables. These latent vectors may be combinations of convenience (i.e., combinations formed for a specific purpose, such as prediction, where the composite is merely a means to an end), or they may be considered to represent real unmeasured variables that underlie the set of observed variables, as is generally the case with the factors resulting from an internal factor analysis, as described in Part III. In either case, the composite is expressed as a linear function of the observed variables.

When we say that $A'$ is a linear function of $P$ and $I$ we mean that it is possible to form an equation of the form

$$A' = b_P P + b_I I \tag{6.1}$$

such that if we were to take a point on the $P$ vector and multiply it by $b_P$, then take a point $I$ and multiply it by $b_I$ and add the two resulting values together (i.e., plot a point at the coordinates given by $b_P P$ and $b_I I$) the resulting value would fall on $A'$. To accomplish this, $b_P$ and $b_I$ are computed in such a way that each reflects that portion of its variable vector that is independent of the other predictor variable vector. Figure 6-2 represents graphically what is going on when we compute these weights. The weights $b_P$ and $b_I$ give the coordinates of $A'$ with respect to these reference axes (which are our observed variables).

It is important to note that the coordinates of $A'$ are expressed as the projections of $A'$ onto the $P$ axis *parallel to* $I$ and onto the $I$ axis parallel to $P$, rather than by the perpendicular projections we have seen before. A moment's thought indicates why this is so. Previously, we have always had independent axes defining our spaces. In such situations, projections perpendicular to one axis (or plane or hyperplane) are actually parallel to the axis whose effect is being omitted. Therefore, in the present case of correlated axes defining our plane we continue to use these parallel projections to provide the independent coordinates of our point.

The reader will also recall from Chapter 2 that the cosine of the angle between two variable vectors adjusted for vector lengths is the correlation them. This property still holds. The cosine of the angle $A'OI$ ($O$ is the origin) is given by the length of the side adjacent to the angle (a segment of $OI$) divided by the length of the hypotenuse ($OA'$). We complete the right triangle by dropping a perpendicular line down to $OI$. If $I$ is expressed in standard score units, then the projection of $A'$ onto $I$ is the correlation between the latent variable $A'$ and the observed variable $I$. Note that there is no restriction that this correlation be independent of other observed variables. It will be what it is regardless of the location of $P$. However, the coordinates that express the location of $A'$ do depend upon the correlation between $I$ and $P$, as is shown in Figure 6-2b, where the $A'I$ correlation is unchanged, but the correlation between $I$ and $P$ (called $P'$) is shown as negative. Where $b_I$ was relatively small in the first figure, it has become quite large. In fact, it may come to exceed 1.0 in some cases. We have much more to say about the correlations between observed variables and latent composites later. For now, or attention focuses on the coordinates of the latent composite in the space defined by the observed variables, the $b$s.

It is important to remember that $P$ and $I$ are expressed as standard scores for the purpose of plotting variables in a space of people. There-

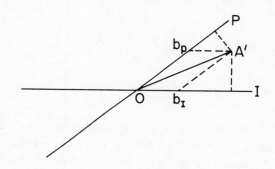

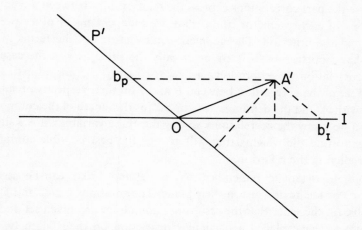

*Figure 6-2  A' may be expressed as a combination of I and P. The coordinates of A' are $b_I$ and $b_p$, and these coordinates are the projections of A' onto I parallel to P and onto P parallel to I, respectively. In part (a), P and I are positively correlated. In part (b) they are negatively correlated.*

fore, equation (6.1) might be better expressed as

$$Z_{A'} = \beta_P Z_P + \beta_I Z_I \qquad (6.2)$$

[It is conventional to use the Greek letter $\beta$ (beta) to represent the co-ordinates of the $A'$ vector in a standard score space, whereas $b$ is used for raw or deviation score spaces.] It can be shown by relatively simple algebra (see McNemar, 1969) that

$$\beta_P = \frac{r_{AP} - r_{AI} r_{PI}}{1 - r_{PI}^2}$$

and

$$\beta_I = \frac{r_{AI} - r_{AP} r_{PI}}{1 - r_{PI}^2} \qquad (6.3)$$

Both these formulas show a striking resemblance to the formulas for their respective part correlations with the criterion, the only difference being the absence of the radical in the denominator. But this is what we should expect. If $r_{AP}$ and $r_{AI}$ are the appropriate weights when $P$ and $I$ are uncorrelated, then something close to the part correlations $r_{A(P \cdot I)}$ and $r_{A(I \cdot P)}$ would be the logical equivalents when $P$ and $I$ are correlated, because the part correlations express the relationships between the unique portions of each predictor (those that are independent of other predictors) and the criterion. The geometric implications of the terms in the right-hand portion of (6.3) are essentially the same as was the case for part correlation. By subtracting $r_{AI} r_{PI}$ (or $r_{AP} r_{PI}$) we are adjusting for the effect of the correlation between $P$ and $I$ on their respective relationships with $A$, and the denominator adjusts for the length of the composite. Thus if we know the correlations among our three variables, it is a simple matter to find the weights that will give us the best possible composite for predicting the criterion.

It is important to remember that the $P$ and $I$ axes can be used to describe or locate *any* line in their plane. The solutions for $\beta_P$ and $\beta_I$ are possible (in the sense that the results are unique for any given set of data) because we have placed a particular restriction on them, namely, that they provide us with a line which has minimum angle with the criterion variable.

**Multiple regression surface.** Now that we have examined the problem of multiple correlation from the point of view of finding a vector that is an optimum composite of our predictor variables when plotted in a space of people, we can approach the matter from the direction of the traditional scatter plot. In this case we have three orthogonal axes which represent our variables, and the people are plotted as points in this three-dimensional space. The scatter plot will have a generally ellipsoidal shape, somewhat like that of a football. For the time being we confine our discussion to the case in which the variables are represented in standard score form so that the center of the ellipsoid will be at the origin (as shown in Figure 6-3), and each of the variables will have unit variance.

In the two-variable case in which people were plotted as points, our problem was to find a regression line about which the sum of squared deviations was as small as possible. When our scatter plot is a three-dimensional one and we have scores for people on two of those dimensions, the task becomes one of finding a *regression plane* or surface from which the points representing the people have the smallest sum of squared deviations. This regression plane has been drawn into the ellipsoid in Figure 6-4. Person $X$, with scores $X_A$, $X_P$, and $X_I$ would actually be plotted at the point marked $X$ in the figure, and person $Y$, whose scores are $Y_A$, $Y_P$, and $Y_I$, would appear at $Y$. However, just as people actually

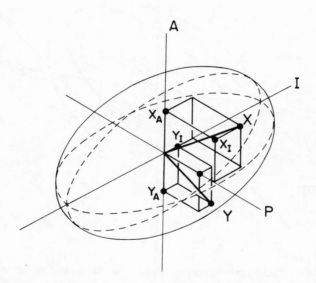

*Figure 6-3   Plot of two people ($X$ and $Y$) in the 3-dimensional scatter plot of scores on variables A, I and P.*

fell at various distances from the regression line in the bivariate case, they also fall at various distances from the regression surface in the multi-variate situation. The point $X$ represents that person's actual standing on $A$ (as well as on $P$ and $I$), but the point $A'_X$ represents his predicted score on $A$. The distance $(A_X - A'_X)$ is the amount of error incurred when we use the regression plane to estimate $A_X$. The problem is to locate the plane in such a way that the sum of squared deviations of points from the plane are a minimum. If there are $N$ individuals, each of whom has an actual $A$ score $(A_i)$ and a score estimated from $P$ and $I$ $(A'_i)$, then we must locate the plane such that

$$\sum_{i=1}^{N} (A_i - A'_i)^2$$

is as small as possible. That is, we locate the plane according to the principle of least squares (which is equivalent to maximum relationship).

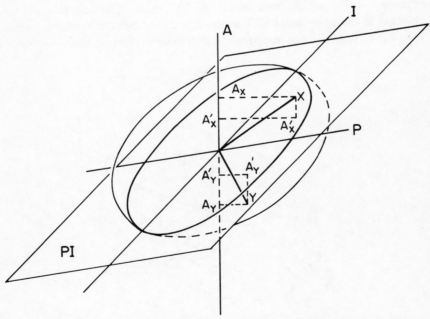

*Figure 6-4   Least-squares regression plane for the scatter plot in Figure 6-3. Points $A_X'$ and $A_Y'$ are predicted A-scores for persons X and Y. Dashed lines project their actual and predicted scores onto the A-axis.*

The regression plane can be defined in much the same way that we defined the regression line. In the bivariate case the required line could be expressed by the point where it crossed the criterion variable axis (its intercept) and the rate of increase in the criterion per unit increase in the predictor (its slope). When dealing with standard scores, the intercept was found to be at the origin (i.e., it was zero) and the slope was given by the correlation coefficient. Because we are dealing with standard scores in our present example, we may safely infer that the regression plane will cross the $A$ axis (criterion axis) at zero. However, because we

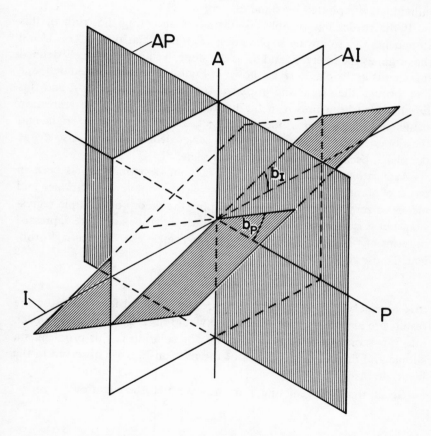

*Figure 6-5 Schematic representation of the regression plane relative to the AI and AP planes. The regression coefficients $b_I$ and $b_p$ are the slopes of the regression plane relative to the I and P axes, respectively.*

are dealing with a 3-space, the slope of the plane is a more complex matter. It can be shown that each of the regression weights (the $\beta$'s) represents the slope of the plane with reference to its particular axis, and that the combination of the two slopes uniquely defines *the* plane about which

$$\sum_{i=1}^{N} (A_i - A'_i)^2$$

is a minimum. That is, $\beta_p$ is the slope that the plane makes with the $P$ axis, and $\beta_I$ is the slope of the plane with respect to the $I$ axis when individuals are plotted by standard scores.

If the reader has trouble visualizing or accepting the truth of this statement, the following explanation may help clarify matters. Using the origin as the intercept and $\beta_P$ as the slope, we have a uniquely defined line that is in the plane $AP$ as shown in Figure 6-5. There is also only one line through the origin and in plane $AI$ that has a slope of $\beta_I$, and this line has also been drawn in the figure. Only one plane, the regression plane we seek, contains both of these lines. Therefore, the origin and the slopes of the lines, which by themselves define the lines, also define the plane, because the lines define the plane.

**Application to raw scores.**   Thus far, our development has been in terms of standard scores, in which the variables have unit variance and means of zero. This is fine for developing the theory of multiple regression, but it is much more convenient to use equations that are expressed in terms of raw scores when applying the principles to a practical problem. To use equation (6.2)

$$Z_{A'} = \beta_P Z_P + \beta_I Z_I$$

it is necessary first to convert all scores to standard form. Because the results are also in standard scores, we would have to convert them back into raw scores to compare $A$ and $A'$. The solution to this problem is a simple and direct generalization of the relationships we observed in the bivariate case.

Recall that for predicting $Y$ from $X$ we used the equation

$$Y' = BX + A$$

where we had

$$B = r_{xy} \frac{S_y}{S_x}$$

and

$$A = \overline{Y} - B\overline{X}$$

Since $r_{xy}$ was the slope of the regression line for standard scores and $B$ was the slope for raw scores, we may substitute the multivariate equivalents and get a raw score multiple regression formula for our example as

$$A' = B_P P + B_I I + \text{intercept}$$

in which

$$B_P = \beta_P \frac{S_A}{S_P} \quad \text{and} \quad B_I = \beta_I \frac{S_A}{S_I}$$

The regression intercept, which is the value of $A$ where the regression plane crosses the $A$ axis, is given by

$$\text{intercept} = \overline{A} - (B_P \overline{P} + B_I \overline{I})$$

We may rephrase these equations into more general terms by substituting $Y$ for the criterion and using $X_1$ and $X_2$ for the predictors. Because the standard score regression coefficients are generally represented by $\beta$, the raw score regression coefficients by the letter $B$, and the intercept by $A$, the above equations become

$$Y' = B_1 X_1 + B_2 X_2 + A \tag{6.4}$$

$$B_1 = \beta_1 \frac{S_Y}{S_{X_1}} \qquad B_2 = \beta_2 \frac{S_Y}{S_{X_2}} \tag{6.5}$$

$$A = \overline{Y} - (B_1 \overline{X}_1 + B_2 \overline{X}_2) \tag{6.6}$$

**The multiple correlation coefficient.** In discussing multiple regression from the point of view of variable vectors plotted in a space of people we said that the problem at hand was to find a composite of our predictor variables that had the smallest possible angle with the criterion vector. From what we know of bivariate correlation it would seem that the angle between the criterion vector and the predictor composite should be related to a correlation of some sort. In fact, the criterion vector $Y'$ (to use our general terminology) is the variable vector of composite scores. If we were to use equation (6.2) to compute a standard score composite score $Z_{Y'}$ for each individual in our sample and plotted this vector in the people space, it would be the composite vector, and each person would have a score on it. The correlation between the criterion variable and the predictor composite is then given by

$$r_{YY'} = \frac{\sum\limits_{i=1}^{N} Z_{Y_i} Z_{Y'_i}}{N}$$

This simple pmc between the criterion and the composite, which is the optimally weighted combination of the predictors, is called the multiple correlation coefficient.

It is not necessary to go to all the trouble to compute the composite score for each individual. Instead, we can use the following formula (see McNemar, 1969 for the derivation):

$$R_{Y \cdot X_1 X_2} = \sqrt{\beta_{x_1} r_{yx_1} + \beta_{x_2} r_{yx_2}} \qquad (6.7)$$

to compute the multiple correlation directly. Here $R_{Y \cdot X_1 X_2}$ is the multiple correlation of $Y$ with the optimum composite of the $X$s. Notice that since the multiple correlation coefficient is defined as a square root, it can never

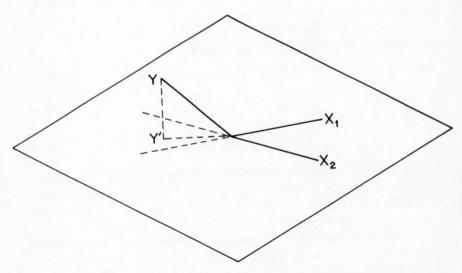

*Figure 6-6   Vector representation showing that the correlation between the criterion variable and the predictor composite is always positive, even when all bivariate predictor-criterion correlations are negative.*

be negative. This does not present a problem because the βs take care of negative bivariate relationships. Even if all the predictors have negative correlations with the criterion, it is possible to find a composite of those predictors that is positively correlated with the criterion, as is shown in Figure 6-6 for two predictors. Note from this illustration that both βs and both predictor-criterion correlations are negative. Thus each of the product terms that enter the multiple correlation would be positive.

**Standard error of estimate.** We have seen that multiple regression analysis results in a regression surface located in such a way as to minimize the sum of squared errors of prediction from the surface. In the bivariate case we were able to develop a formula, called the standard error of estimate, to reflect the average of these squared errors. We can do the same thing with respect to our criterion variable in the case of multiple regression. The definition of the standard error of estimate for multiple correlation is

$$S_{Y \cdot X_1 X_2} = \sqrt{\frac{\sum_{i=1}^{N} (Y_i - Y_i')^2}{N}}$$

However, the squared multiple correlation coefficient is equal to the proportion of variance in $Y$ that is related to or explained by the set of predictors. Therefore, $1 - R^2_{Y \cdot X_1 X_2}$ is the proportion of criterion variance that is unexplained by the predictors and

$$S_{Y \cdot X_1 X_2} = S_Y \sqrt{1 - R^2_{Y \cdot X_1 X_2}} \tag{6.8}$$

is a convenient formula for the standard error of estimate. This equation gives us the standard deviation of deviations of the criterion scores from the regression plane directly. The companion formula

$$S_{Y'}^2 = S_Y^2 R^2_{Y \cdot X_1 X_2}$$

gives us the variance of the predicted criterion scores. Thus it is easy to see that $R^2_{Y \cdot X_1 X_2}$ is the proportion of variance in the original $Y$ scores which is related to the set of $X$ scores (see McNemar, 1969 for derivations) because $R^2 = S_{Y'}^2 / S_Y^2$. It is also easy to see that if we have $Y$ in standard score form ($S_Y^2 = 1$), $R^2$ is the proportion of $Y$ variance that is predictable. This means that the multiple correlation coefficient itself is the standard deviation of composite scores and the length of the composite in the standard score space of people.

There is another interesting feature of the relationship between the predictors and the criterion that can be developed from the standard score prediction equation and which should come as no surprise to the reader, given our observation of the relationship between the $\beta$ weights and part correlations. It can be shown that the variance in predicted standard scores is

$$S^2_{Z'_Y} = \beta^2_{X_1} + \beta^2_{X_2} + 2\beta_{X_1}\beta_{X_2}r_{X_1 X_2}$$

in which the last term is a covariance term reflecting the fact that the predictors are correlated. The important feature of this equation is that it tells us that the squared $\beta$ weights reflect the *relative* importance of the $X$ variables in the prediction equation. The larger a variable's weight, the greater is the importance of that variable in the accurate prediction of the criterion. The importance of this aspect of $\beta$ weights will become more clear when we discuss the topic of stepwise regression. (Note that a $\beta^2$ is not a proportion of variance, but a relative contribution. In this sense, it is different from a correlation coefficient.)

## MULTIPLE DESCRIPTION

This far in this chapter we have discussed the more common application of multiple correlation, its use as a predictive tool. There is a second use to which the procedures of multiple correlation can be put, but which has seldom, if ever, appeared in the literature. The ideas developed in this section are based on a paper by Bartlett (1948) in which he discussed the distinction between canonical analysis and traditional factor analysis. The distinction he made has not been widely discussed, but it provides a very useful way of categorizing the multivariate procedures that are primarily correlational in nature.

The distinction Bartlett drew was based on a recognition of the fact that canonical analysis and factor analysis are both in a sense descriptive procedures. Both provide ways for describing, by means of composites, the relationships among the variables in a set of variables being analyzed. They differ in the criteria by which the composites are developed. In traditional factor analysis there is a single set of variables, and the composites are developed in such a way as to maximize the proportion of the variance of these variables that is "explained" by each successive composite. Bartlett called these procedures "internal factor analysis." Part III is devoted to an explanation of various methods by which this general goal is reached in this widely used multivariate procedure.

In canonical analysis, of which multiple regression is a special case, the problem is somewhat different. Because we are now familiar with multiple regression, we will use it at this point to describe what Bartlett called "external factor analysis." In the next chapter we shall see how the ideas described below generalize to the canonical analysis situation.

When we have a set of $X$ variables (we have been calling them predictors, but when we wish to describe rather than to predict we can refer to them simply as the $X$ set) and a $Y$ variable, we have seen that it is possible to develop a composite of the $X$ set that is maximally related to $Y$. The composite of the $X$ set is developed under the restriction of maximum relationship to a variable that is external to the set. Because the composite of the $X$ set is developed subject to maximizing its relationship with an external variable, this class of procedures is given the name *external factor analysis*, which may be contrasted with the criterion of maximizing within-set variance that characterizes internal factor analysis. The general term *factor analysis* is used to refer to any procedure that involves the development of a composite of variables for description.

In the multiple regression solution developed earlier in this chapter we saw that the $\beta$ weights reflected the relative contribution of each variable in the $X$ set to that composite of the $X$ variables which had the maximum correlation with $Y$. We also saw that these same $\beta$ weights were related to the part correlations of the $X$ variables with the $Y$ variable, in the sense that they provided an index of the degree of relationship between each $X$ and the $Y$ that was independent of the other $X$. However, there may be occasions in which we wish to describe the composite of the $X$s in terms of its relationship with each of the separate $X$ variables. Description in this sense would require knowing the correlation between each variable and the composite, and nothing we have developed thus far provides us with this information.

It might be argued that the $\beta$ weights provide the required information, but such is not the case. The $\beta$ weights give us the relative contribution of each variable to the variance of the composite, but to the extent that the $X$ variables are correlated, the one with the larger correlation with the criterion will receive a large weight at the expense of the other. Graphically, the most extreme example would be as shown in Figure 6-7. This figure has been drawn to show $X_1$ and $X_2$ moderately correlated, but $Y$ correlated with both in such a way that it projects directly onto $X_1$. In this case $X_1$ would have a large weight, and the weight of $X_2$, reflecting its independent contribution to the composite, would be zero. However, $X_2$ does not have a zero correlation with the composite, meaning that a description of the composite in terms of the $\beta$ weights would give a false impression of the relationship. The correlations of the $X$ variables with the composite would accurately reflect the fact that both $X_1$ and $X_2$ are correlated with the composite, even though $X_2$ makes no contribution to the variance of the composite. Admittedly, this is an extreme example, but the phenomenon described above does occur and with greater frequency as the number of variables in the $X$ set increases. Therefore, when description of the composite of $X$ variables is desired, it is necessary to compute the correlations of the $X$ variables with the composite. As described earlier, these correlations are the perpendicular projections of the composite vector onto the variable vectors.

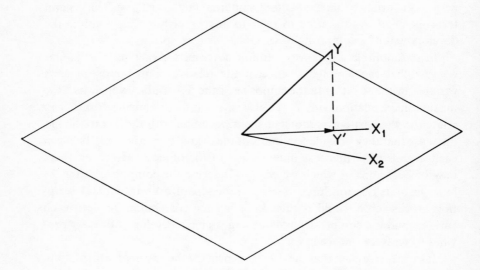

*Figure 6-7   Vector diagram of the case where $X_2$ has a $\beta$-weight of zero and a substantial positive correlation with $Y'$. $Y'$ falls exactly on $X_1$ and has a correlation of 1.0 with $X_1$.*

Computation of the required correlations may be accomplished in either of two ways. The most obvious and straightforward of these is to compute each person's score on the composite by using equation (6.2), or its raw score equivalent (6.4), and then to compute the correlations of these composite scores with the appropriate sets of scores on the original $X$ variables. However, a much simpler shortcut accomplishes the same result. Specifically, it can be shown (see Cooley and Lohnes, 1971) that the correlations of scores on the predictor variables ($X_i$s) with scores on the composite ($Y'$) that is the best predictor of the criterion ($Y$) are given by

$$r_{X_i Y'} = \frac{r_{X_i Y}}{R_{Y \cdot X_1 X_2 \cdots}} \tag{6.9}$$

Geometrically, what is going on is that, once again, the denominator term is correcting for the length of the vector. Consider Figure 6-7. The vector $Y$ defines a plane that is perpendicular to the plane of $X_1$ and $X_2$. There is one and only one plane in the 3-space of $X_1$, $X_2$, and $Y$ that is

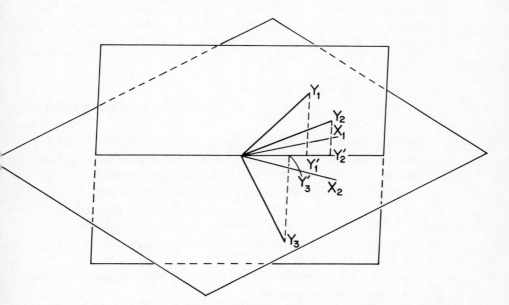

*Figure 6-8   The correlations of $Y'$ with $X_1$ and $X_2$ are functions of the location of $Y'$ relative to $X_1$ and $X_2$. These correlations or component loadings are not affected by the magnitude of the multiple correlation.*

perpendicular to the $X_1$, $X_2$ plane and contains $Y$. The intersection of these two planes defines the direction of $Y'$, and $R_{Y \cdot X_1 X_2}$ determines its length. The one exception to this statement occurs when $Y$ is perpendicular to the $X_1$, $X_2$ plane, in which case $R_{Y \cdot X_1 X_2}$ is zero and the $Y'$ vector does not exist.

Now, let us consider for a moment what would happen if we were to take the location of the $Y$ plane just described as fixed and let the $Y$ vector vary. This hypothetical situation is shown in Figure 6-8. We must require that $Y$ remain in the positive half of the plane; otherwise the direction of $Y'$ would change. However, within this broad restriction we can move $Y$ up and down at will. Remembering that $Y$ remains the same length because we have standard scores, the effect of movement of $Y$ on $Y'$ is obvious. As Y becomes more nearly perpendicular to the plane of $X_1$ and $X_2$, $Y'$ will become shorter. The angles between the $X$ variables and $Y'$ will remain unchanged, even though the angles between the $X$s and $Y$ become greater or smaller. Thus, given the location of the $Y$ plane, which is a function of the *relative* magnitudes of $r_{x_1 y}$ and $r_{x_2 y}$, the correlation of $X_1$ and $X_2$ with $Y'$ are constant, regardless of the magnitude of the multiple correlation! The most obvious example of this concept is shown in Figure 6-7, where the correlation between $Y'$ and $X_1$ remains 1.00 so long as $r_{x_1 y}$ is positive.

## A THREE-VARIABLE EXAMPLE

To illustrate the computations and results of a multiple correlation analysis, we use the data from Table 5-1, which are reproduced in Table 6-1 for ease of explanation. In this example we will determine the regression equation for predicting achievement $(Y)$ from potential $(X_1)$ and interest $(X_2)$, assuming that the means of the three variables are 45, 57, and 70 and that the standard deviations of the three variables are 10, 12, and 15, respectively. All the computational details are included in Table 6-1.

We can use the information in the table to specify the standard score and raw score regression equations

$$Z'_Y = .53 Z_{X_1} + .24 Z_{X_2}$$
$$Y' = .44 X_1 + .16 X_2 - 8.72$$

The multiple correlation coefficient is .644, which is only slightly higher than the best bivariate correlation, $r_{AP}$. This would lead us to conclude that interest adds very little to potential as a predictor of achievement. However, if we consider the squared correlations we find that potential accounts for 36% of the variance in achievement, whereas the best combination of potential and interest accounts for 41.4% of the achievement score variance. Interest, when used in the composite, permits us to account for an additional 5.4% of the $A$ variance. Later we discuss significance tests for gains by adding variables to a multiple correlation, but the 5.4% increase might be considered promising. From the standard error of

**TABLE 6-1**

*Computation of the Multiple Correlation and Regression Solution for Three Variables*

|   | $A$ | $P$ | $I$ |
|---|-----|-----|-----|
| $A$ | 1.00 | | |
| $P$ | .60 | 1.00 | |
| $I$ | .40 | .30 | 1.00 |

$$\beta_{x_1} = \frac{.6 - .4(.3)}{1 - .3^2} = \frac{.48}{.91} = .53$$

$$\beta_{x_2} = \frac{.4 - .6(.3)}{1 - .3^2} = \frac{.22}{.91} = .24$$

$$B_{x_1} = .53\left(\frac{10}{12}\right) = .53(.83) = .44$$

$$B_{x_2} = .24\left(\frac{10}{15}\right) = .24(.667) = .16$$

$$A = 45 - [.44(57) + .16(70)] = 45 - (25.08 + 11.2) = -8.72$$

$$R_{y \cdot x_1 x_2} = \sqrt{.53(.6) + .24(.4)} = \sqrt{.414} = .644$$

$$R^2 = .414$$

$$S_{y \cdot x_1 x_2} = 10\sqrt{1 - .644^2} = 10\sqrt{.586} = 10(.765) = 7.65$$

$$r_{x_1 Y'} = \frac{r_{x_1 Y}}{R} = \frac{.6}{.644} = .93$$

$$r_{x_2 Y'} = \frac{r_{x_2 Y}}{R} = \frac{.4}{.644} = .62$$

estimate we can conclude that on the average our predictions will be in error by about 7.65 units.

The information in the last section of the table is particularly interesting in that it provides the description of the composite. Notice that there are substantial differences between the $\beta$ weights and the correlations of the variables with the composite (later an argument is made that these should be called component loadings). Variable $X_1$ correlates .93 with the composite, whereas $X_2$ has a correlation of .62 with the composite.

## MORE THAN TWO PREDICTORS

Seldom does an investigator have a multiple correlation problem as small and simple as the one we have discussed. It is more common to have 10 variables than two. For these more complex problems there are only two changes that must be made in what was said in the previous sections of this chapter. The first is in the nature of the geometric representation. It is necessary to have one dimension for each variable, which means that we must start thinking in hyperspaces and cannot use diagrams very well.

The second change that must be made is in the way in which the $\beta$ weights are computed. Once there are more than two predictors, it becomes very difficult to compute the $\beta$s by equations such as (6.3). Instead, we can use matrix algebra to arrive at a simple equation that can be readily solved by a computer. If we say that $\mathbf{R}$ is a matrix of correlations among the predictor variables and that $\mathbf{c}$ is a vector of correlations of the predictors with the criterion, then the vector of $\beta$ weights, $\boldsymbol{\beta}$ is given by the equation (see Cooley and Lohnes, 1971 for the derivation):

$$\boldsymbol{\beta} = \mathbf{R}^{-1} \cdot \mathbf{c} \qquad (6.10)$$

Because virtually all computer installations have prepackaged routines for finding the inverse of a matrix (indeed, it is hard to imagine one without a multiple correlation program), the computation of $\beta$ weights by this method requires relatively little labor.

Once the vector of $\beta$s is available, the raw score regression weights can be obtained by multiplying each $\beta$ weight in turn by the ratio of the standard deviation of the criterion variable to the standard deviation of the predictor whose $\beta$ weight is being considered. The standard score regression equation for $m$ predictors

$$Z_Y' = \beta_{X_1}Z_{X_1} + \beta_{X_2}Z_{X_2} + \beta_{X_3}Z_{X_3} + \cdots + \beta_{X_m}Z_{X_m} \quad (6.11)$$

yields the raw score regression equation

$$Y' = \beta_{X_1}\left(\frac{S_Y}{S_{X_1}}\right)X_1 + \beta_{X_2}\left(\frac{S_Y}{S_{X_2}}\right)X_3 + \beta_{X_3}\left(\frac{S_Y}{S_{X_3}}\right)X_3 + \cdots + \beta_{X_m}\left(\frac{S_Y}{S_{X_m}}\right)X_m + A$$

or

$$Y' = B_{X_1}X_1 + B_{X_2}X_2 + \cdots + B_{X_m}X_m + A \quad (6.12)$$

where the value of $A$ is given by

$$A = \bar{Y} - [B_{X_1}\bar{X}_1 + B_{X_2}\bar{X}_2 + B_{X_3}\bar{X}_3 + \cdots + B_{X_m}\bar{X}_m] \quad (6.13)$$

That is, every principle from the two-predictor case generalizes directly to problems involving more than two predictors, including the multiple correlation, which is given by

$$R_{Y \cdot X_1 X_2 \cdots X_m} = \sqrt{\beta_{X_1}r_{YX_1} + \beta_{X_2}r_{YX_2} + \cdots + \beta_{X_m}r_{YX_m}} \quad (6.14)$$

The standard error of estimate, $S_{Y \cdot X_1 X_2 \cdots X_m} = S_Y\sqrt{1 - R^2}$, has the same meaning as given above except that now it involves deviations of actual $Y$ scores from the predictor hyperplane. The $R^2$ may be interpreted as the proportion of variance in $Y$ scores that is accounted for by the composite of the $m$ $X$ scores, and the component loadings or correlations of the predictors with the composite may be used to describe the composite in terms of its correlations with the predictors.

**Stepwise multiplication correlation.** One of the things often found when a large number of predictors are used in a multiple correlation analysis is that a few variables receive substantial $\beta$ weights and that the rest of the $\beta$s are near zero. Even when all the predictors have moderate to large correlations with the criterion, the correlations among the predictors prevent most of them from having much effect on the predictor composite. When this situation occurs there are three reasons why most researchers are well advised to look for the subset of predictors that will accomplish most of the prediction and to discard the rest of the variables, a procedure known as stepwise correlation.

The first reason for seeking an optimum subset of the predictors is purely economic. When working in an applied situation, for example, a personnel placement office, in which a large number of predictions are to be made, it is economically unwise to gather information on variables that will not contribute to the accuracy of prediction. There is no reason to gather information on 10 tests when the best 5 of them will predict just as accurately as the full set of 10. The problem, which we address shortly, is how to decide which variables constitute an optimum subset.

A second reason for using the minimum number of variables is simplicity of description. This problem, which has received a great deal of attention from factor analysts under the heading of simple structure, is an appropriate concern when description is a goal of a multiple correlation analysis. It is generally easier to justify an interpretation of a predictor composite when only a few variables are involved than when there are a large number of component loadings to be considered. A stepwise approach may provide this simplicity under the restriction of maximum relationship with the criterion.

A final argument for the use of stepwise procedures in multiple correlation analysis can be made in terms of the stability of the resulting regression equation. It is in the nature of all least-squares procedures to capitalize on relationships between variables that are the result of chance variations in the sample obtained, rather than being characteristic of the population from which the sample was taken. The larger the number of variables, the greater is the possibility that the regression equation will reflect chance relationships. When these chance relationships do affect the $\beta$ weights, the regression equation will not yield accurate predictions for a new sample. In an extreme case it is possible that the predictions made from the regression equation will be less accurate than random guesses. By reducing the number of variables in the regression equation, we reduce the likelihood that chance relationships in the sample will have a substantial affect on the $\beta$ weights.

There are two ways to proceed in a stepwise analysis. On the one hand, we may start with the best single predictor and add one variable at a time until the addition of another variable raises the multiple correlation but little. To use this approach we must calculate the multiple correlation of the criterion and the best predictor when it is paired with each of the other predictors in turn, which requires the computation of $m - 1$ multiple correlations when there are $m$ predictors. Next, each of the remaining variables is added in turn to the best pair of predictors, and the $m - 2$ multiple correlations are computed. This procedure is repeated until there is no variable that adds significantly to the multiple correlation of the already selected variables with the criterion. This is called a step-up approach.

Alternatively, we may begin with the multiple correlation of all predictors with the criterion. With this complete analysis in hand we may drop the variable that has the smallest $\beta$ weight and recompute the multiple correlation to determine whether there is a significant loss in predictive accuracy. This process is repeated, each time dropping the variable with the smallest $\beta$ weight, until there is a significant drop in the multiple correlation. When a significant drop occurs, the solution preceding the drop is taken as the final regression equation. This is called the step-down approach, and it will not necessarily give the same results that are obtained from a step-up procedure. Perhaps the best results can be obtained by using both approaches and selecting the equation that is most in line with the investigator's objectives.

**Tests of significance.** There are two tests of statistical significance that are important in multiple correlation analysis. First, we may wish to know whether a particular multiple correlation is significantly greater than zero. This is a test of the probability that our obtained $R$ is a result of chance factors in sampling from a population in which the true $R$ is zero. The proper test of the hypothesis that the obtained correlation is a random fluctuation from a population value of zero is given by

$$F = \frac{R^2/m}{(1 - R^2)/(N - m - 1)} \tag{6.15}$$

in which $F$ has $m$ numerator degrees of freedom and $(N - m - 1)$ denominator degrees of freedom. If the probability of this $F$ is less than, say, .01 we may safely conclude that the multiple correlation of the set of predictors with the criterion is not zero in the population from which our sample is drawn.

The other test of significance that is useful in multiple correlation analysis tests whether a particular variable makes a worthwhile contribution to the accuracy of prediction. Actually, this amounts to testing the significance of a $\beta$ weight. It also provides a method of determining whether the $R$ obtained by using a set of $(m + p)$ variables is enough greater than that obtained by using just $m$ of them to justify using the $p$ extra variables. (Of course, the answer is given solely in terms of statistical significance and probability. Economic factors of test cost, testing time, etc., may lead to the conclusion that the gain in predictive accuracy is not worth the added cost.)

If we say that $R_1$ is the multiple correlation obtained when $(m + p)$ predictors are used and that $R_2$ is the multiple correlation found when only $m$ of them are retained, then

$$F = \frac{(R_1{}^2 - R_2{}^2)/p}{(1 - R_1{}^2)/[N - (m + p) - 1]} \qquad (6.16)$$

provides a test of whether $R_1$ is significantly greater than $R_2$. This $F$ has $p$ and $[N - (m + p) - 1]$ degrees of freedom. When $p = 1$, this is a test of whether the addition of a single variable provides a significant increase in the multiple correlation and is therefore a test of the significance of the $\beta$ weight of that variable. Although testing all of the $\beta$s in a regression equation by the procedure outlined above seems like a large order, it is relatively simple to do with a computer. Also, a procedure like this one might be used to test the significance of higher-order part correlations because of the relationship between part correlations and $\beta$ weights.

**Cross-validation.** Earlier in this chapter it was noted that least-squares procedures capitalize on all relationships that exist in a particular sample. It is a fact of life for the investigator who would predict behavior that successive random samples from a population will differ in the nature and extent of the relationships among some of the variables being studied. We use the term *sample-specific covariation* to refer to relationships that are found in one random sample from a population but not in other random samples from the same population.

When only two variables are involved, sample-specific covariation results in the sampling distribution of the product moment correlation, and as such the limits of its effects can be judged by a confidence interval on $r$. However, when several variables enter a regression equation the problem becomes much more complex. It is reasonably safe to say that the regression equation based on a particular sample will not work as well when applied to a new sample as it did for the sample on which it was developed, but it is hard to estimate beforehand what the effects of sample-specific covariation will be other than to conclude that the $R$ will shrink.

McNemar (1969, p. 205) provides a formula by which we can estimate the degree of shrinkage when we apply our regression equation to a new sample. However, this formula

$$\tilde{R}^2 = 1 - (1 - R^2)\frac{N - 1}{N - m} \tag{6.17}$$

in which $\tilde{R}^2$ is the "shrunken" $R$, $N$ is sample size, and $m$ is the number of variables, was derived under an assumption of ideal random samples. Under the more realistic conditions of less-than-perfect sampling this formula gives us an estimate of the minimum amount of shrinkage we can expect. With reference to our previous comments about the stability of regression equations and the use of stepwise procedures, note that as the number of variables becomes small relative to sample size, we would expect to experience less shrinkage of our multiple correlation. Under certain conditions discussed by McNemar, the estimate of the shrunken $R$ may yield negative values. This is most likely to happen when the number of variables is large relative to the sample size.

When working on a practical problem, for example the prediction of occupational or academic success, the investigator is not really interested in the accuracy of prediction in the sample at hand. Rather, he is concerned with how well the regression equation from his present sample will work for a new group of individuals. It is not his goal to predict criterion scores for those individuals on whom information about criterion performance is already available; rather, it is to be able to make the best possible estimate of criterion performance for new individuals at a time well before their criterion scores will be available. At this point the question is how well the regression equation will predict for a new group, and this question must be answered empirically by trying out the equation.

If we consider the original sample, called the *development sample*, to be a sample from a given population that has a population regression equation, then the $\beta$ weights found in the development sample may be considered to be estimates of the unknown (and unknowable) population $\beta$ weights. When the development sample is a random sample from the population, the sample $\beta$ weights are unbiased estimators of the population $\beta$ weights, and we should experience relatively little loss in predictive accuracy when applying our regression equation to a new sample from the population. The procedure of trying the regression equation out on a new sample to determine the stability of the equation has been given the name *cross-validation*, a name that reflecting the fact that the procedure was originally devised to determine the validity of scoring keys in which differential weights are given to the items of a test or inventory.

The need for cross-validation has been adequately demonstrated in several studies. Cureton (1950) has shown that it is possible to find weights that will result in good prediction in the development sample when the data are essentially random numbers. When this prediction equation was applied to a cross-validation sample the accuracy of prediction was zero, as it should have been. The weights were based entirely on sample-specific covariation, and Cureton concluded that no confidence can be placed in a set of weights unless they have been shown to yield accurate prediction in cross-validation.

To carry out a cross-validation study it is necessary to have predictor and criterion information on the cross-validation sample as well as on the development sample. In effect, this means that the original sample must be broken down into two parts or that two separate data collections must be performed. There are arguments for both approaches. On the one hand, collecting all the data at one time is generally more convenient and less expensive. However, there may be situational or temporal factors that increase sample-specific covariation in both samples when all of the data are collected at one time. That is, splitting one large sample into development and cross-validation samples may yield a regression equation in which the same sample-specific covariation is present in both samples, a situation that will tend to give an inflated estimate of the accuracy of prediction for new samples. Separate data collections for development and cross-validation samples, although uncommon in practice, will give a better estimate of the accuracy that can be expected from the regression equation when it is put into use.

Three strategies have been proposed for splitting a sample for development and cross-validation. The first of these, historically, is the *holdout group method*, in which a relatively small number of individuals, usually about 25%, are set aside as the cross-validation group; the others are used to develop the regression equation. After development, the regression equation is applied to the data from the holdout group to determine whether the equation will cross-validate at an acceptable level. This procedure is most justifiable when the number of individuals available is relatively small.

The holdout group method has been criticized because it wastes data. Mosier (1951) noted that one would like to base the regression equation on as many individuals as possible to obtain the best estimate of the weights. He noted that the data for the cross-validation group are lost for the purposes of equation development if the holdout group method is used. As an alternative, he proposed the method of *double cross-validation* (note the location of the hyphen) in which the available sample is split in half and two separate regression equations are developed. Each regression equation is then cross-validated on the half of the original sample on which it was not developed. He suggested that when both equations proved satisfactory in cross-validation and the weights from the two sets of data were similar, the final equation could be based on the combined data. The investigator could then be reasonably sure that the final equation would cross-validate and that the weights would be based on the largest possible sample, thereby providing better estimates of the population values.

A modification of the double cross-validation procedure has been proposed by Norman (1965), who gave it the name *double split cross-validation*. This approach has not been used very frequently. It requires splitting the original sample into four parts, one for item selection, two for equation development, and one for cross-validation. Because it was designed for instrument development and has not been widely used, we do not consider it further.

We have discussed cross-validation as a means by which the stability of the regression equation may be determined. This is a problem that concerns the stability of the $\beta$ weights and is of primary concern when prediction is the goal of a study. In prediction research, cross-validation must be considered a mandatory part of the research design. However, there are two other questions that may be asked of the data by using multiple correlation. The investigator may wish to obtain a description of the predictor composite without wanting to actually use the composite for prediction. When multiple correlation analysis is used in this way it is an external factor analysis. The function of cross-validation in this context is discussed in the next chapter.

The other question that may be posed in a multiple correlation analysis is simply the degree of relationship which exists in the population between the predictors and the criterion. In this case the object of the study is to estimate as accurately as possible the multiple correlation coefficient in the population. The best estimate of the population $R$ is obtained when that $R$ is computed from all available cases. However, this question is seldom of interest in the absence of one of the other two. When neither a replication (a second development analysis on a new set of data) nor a cross-validation has been performed, we are very limited in what we can conclude from our study. The investigator must be cautious and not attempt to make generalizations about the regression equation or the component loadings unless an appropriate cross-validation or replication has been performed, lest his results be deserving of Cureton's comment about uncross-validated regression equations—baloney!

## SUMMARY

In circumstances in which we have more than one predictor variable and we wish to get the best possible prediction of a criterion, it is necessary to form a composite variable. This composite, which is a latent variable and not directly measured, is formed by using differential and optimum weights which are applied to the directly measured predictor variables. The correlation between the composite variable and the criterion variable is the multiple correlation coefficient, the square of which is the proportion of variance of the criterion variable that is accounted for by the composition of the predictor set.

In a standardized space of people the composite variable may be viewed as a vector that is completely within the subspace defined by the predictors and which has the smallest possible angle with the criterion vector. The $\beta$ weights, which are the coordinates of the composite vector, reflect the relative importance of the predictor variables in determining the composite. The location of the composite vector relative to the predictor vectors is given by the component loadings, which are the correlations of the predictors with the composite and provide an external factor analysis of the set of predictor variables under the restriction of maximum relationship with the criterion.

In a space of standardized variables the $\beta$ weights give the slope of the regression surface relative to the axes that represent the variables. The $\beta$ weights can be converted into raw score regression weights by multiplying each $\beta$ by the ratio of the criterion standard deviation to the standard deviation of that predictor, in which case the intercept is the point on the criterion variable axis where the regression surface crosses that axis. The location of the regression surface is defined in such a way that the average of the sum of squared errors of prediction from the surface is a minimum and is given by

$$\sqrt{\Sigma(Y - Y')^2/N} = S_Y\sqrt{1 - R^2}$$

When a fairly large number of predictors are available, it is often advisable to use either a step-up or a step-down procedure to select a best possible subset of predictors. This approach yields a more stable regression equation that is less subject to shrinkage and simplifies description of the composite.

The use of multiple correlation analysis for prediction of criterion performance in a new group requires cross-validation of the regression equation to evaluate the stability of the $\beta$ weights. A holdout sample may be used, but it is generally preferable to use double cross-validation when possible. An external factor analysis requires cross-validation of the results to confirm their generality. The possibility that the variables retained by a stepwise procedure may be those that show the largest amounts of sample-specific covariance makes it doubly necessary to cross-validate equations developed in this way.

## AN EXAMPLE

The example we will use for the remaining topics in Part II comes from some industrial psychological data collected by Dr. John Sauer. The variables, their means, and standard deviations for a sample of 130 men are given in Table 6-2. The first 18 variables are characteristics of the 130 individuals' backgrounds and may be considered predictors; variables A to D provide job-related information and are potential criterion variables. We will use the 18 background variables to predict salary. Before any additional analyses were performed, standard scores were computed for all individuals using the total group means and standard deviations, and the group was divided semi-randomly (see the example in Chapter 8) into a group of 100 for development and a group of 30 for cross-validation. The correlations necessary for the development analysis (based on the 100 individuals in the development group) are given in

**TABLE 6-2**

*Means and Standard Deviations for 18 Biographical and 4
Occupational Variables.*

| Variable Name | Mean | SD |
|---|---|---|
| 1. Length of residence in state | 100.39 | 114.90 |
| 2. Age | 30.34 | 6.89 |
| 3. Years of education | 15.88 | 1.53 |
| 4. High school GPA | 2.50 | .73 |
| 5. High school size | 290.62 | 309.01 |
| 6. Class standing | 28.89 | 19.64 |
| 7. Number HS offices held | 1.84 | .86 |
| 8. Years of college | 3.89 | 1.66 |
| 9. Time on subsequent courses | 5.82 | 13.76 |
| 10. Home owned | 1.61 | .49 |
| 11. Time at present residence | 33.30 | 48.74 |
| 12. Number of children | 3.20 | 1.54 |
| 13. Birth order | 1.72 | 1.18 |
| 14. Weight | 177.09 | 19.45 |
| 15. Height | 71.26 | 2.34 |
| 16. Number of organizations joined | 1.74 | 1.80 |
| 17. Ultimate salary desired | 45.29 | 36.28 |
| 18. Level of aspiration | 1.29 | .48 |
| A. Level of occupation | 2.60 | .96 |
| B. Tenure | 46.17 | 64.95 |
| C. Salary | 10.92 | 7.29 |
| D. Supervisory | 1.00 | .87 |

Table 6-3. Note that decimal points have been omitted and that only the lower triangle of the correlation matrix is presented.

Application of equation (6.10) yields the $\beta$ weights given in the next-to-last row of the table. From these weights we see that age makes the largest independent contribution to the composite that is maximally related to salary. The fact that the weight is positive indicates that older men tend to earn higher salaries, a finding which is not surprising. On the other hand, years of college education has a negative weight, meaning that those with more education, other things being equal, tend to earn lower salaries! Other $\beta$ weights of modest size indicate that higher high school GPA and class standing, longer time at present residence, being a member of more organizations, and having higher ultimate salary desired, all

**TABLE 6-3**

*Correlations among 18 Biographical Variables, their Correlations with Salary, and their Regression Weights and Component Loadings.*

| | 1 | 2 | 3 | 4 | 5 | 6 | 7 | 8 | 9 | 10 | 11 | 12 | 13 | 14 | 15 | 16 | 17 | 18 | C |
|---|---|---|---|---|---|---|---|---|---|---|---|---|---|---|---|---|---|---|---|
| 1 | 1.00 | | | | | | | | | | | | | | | | | | |
| 2 | 02 | 1.00 | | | | | | | | | | | | | | | | | |
| 3 | −06 | 02 | 1.00 | | | | | | | | | | | | | | | | |
| 4 | 00 | 13 | 12 | 1.00 | | | | | | | | | | | | | | | |
| 5 | −09 | −16 | 03 | −13 | 1.00 | | | | | | | | | | | | | | |
| 6 | 05 | 00 | −07 | −16 | 04 | 1.00 | | | | | | | | | | | | | |
| 7 | −10 | −19 | 19 | 12 | −02 | −06 | 1.00 | | | | | | | | | | | | |
| 8 | −18 | 18 | 75 | 11 | −02 | 00 | 21 | 1.00 | | | | | | | | | | | |
| 9 | 00 | 13 | 13 | 17 | 00 | 04 | 00 | 12 | 1.00 | | | | | | | | | | |
| 10 | −07 | −47 | 00 | 04 | 18 | 00 | 11 | −10 | 11 | 1.00 | | | | | | | | | |
| 11 | 22 | 56 | −06 | 22 | −15 | 08 | −06 | 05 | 01 | −35 | 1.00 | | | | | | | | |
| 12 | 02 | −05 | −05 | 03 | −18 | −09 | 01 | −03 | −06 | 02 | −06 | 1.00 | | | | | | | |
| 13 | −05 | −11 | 09 | −05 | −01 | −05 | 06 | −04 | −01 | −15 | 06 | 35 | 1.00 | | | | | | |
| 14 | 14 | −02 | −12 | 00 | 01 | 05 | 04 | −16 | −05 | −14 | 05 | 11 | 17 | 1.00 | | | | | |
| 15 | −03 | −26 | −06 | 07 | 10 | −02 | −03 | −13 | 03 | 09 | −12 | 16 | 09 | 65 | 1.00 | | | | |
| 16 | 05 | 45 | 22 | 17 | −09 | 02 | 07 | 16 | 16 | −31 | 34 | −06 | 09 | 17 | −14 | 1.00 | | | |
| 17 | −15 | −20 | 01 | 12 | 00 | −06 | 07 | −08 | 08 | 12 | −09 | 07 | 16 | 09 | 19 | −01 | 1.00 | | |
| 18 | 00 | −21 | 12 | −02 | 19 | −01 | 06 | 02 | 09 | 19 | −03 | 15 | 23 | 22 | 33 | 02 | 43 | 1.00 | |
| C | 09 | 38 | 00 | 21 | −01 | 10 | −04 | −04 | 12 | −24 | 38 | 00 | 08 | 11 | 01 | 35 | 13 | 10 | 1.00 |
| βs | 04 | 27 | 08 | 13 | 08 | 12 | 03 | −19 | 03 | −10 | 15 | 02 | 02 | −05 | 07 | 16 | 14 | 08 | |
| Bs | 00 | 29 | 40 | 1.28 | 00 | 04 | 24 | −85 | 02 | −1.52 | 02 | 08 | 14 | −02 | 63 | 03 | 03 | 1.20 | $A = -22.34$ |
| $r_{xy'}$ | 16 | 68 | 00 | 38 | −02 | 18 | −07 | −07 | 21 | −43 | 68 | 00 | 14 | 20 | 02 | 62 | 23 | 18 | |

indicate a higher present salary in that they make positive contributions to the predictor composite.

The accuracy with which the composite defined by the $\beta$ weights in Table 6-3 predicts salary is given by the multiple correlation, which, using equation (6.14), is

$$R_{Y \cdot X_1 \cdots X_{18}} = \sqrt{.04(.09) + .27(.38) + \cdots + .08(.10)} = .56$$

and is statistically significant ($F = 2.06$; $p < .05$). The 18 predictor variables are, therefore able to account for 31% ($= R^2$) of the variance in salary, and the standard error of estimate is found by equation (6.8) to be 6.06.

As we noted earlier, one would ordinarily not use equation (6.11) for a practical prediction problem, because it requires calculation of standard scores. Rather, one of the forms of (6.12) would be used. The raw score regression weights ($\beta$s) for use in this formula are given in the next line of Table 6-3 in our example. The B weights do not tell us much about the relative contributions of the different variables to the prediction of salary, because predictor variables with very small standard deviations will receive quite large B weights even when their $\beta$ weights are modest. Variables 4, 8, 10, and 18 in our present example illustrate this feature. All these variables have substantially lower $\beta$ weights and correlations with salary than does the age variable. However, their B weights are much higher because they have small standard deviations. Thus the B weights are convenient for prediction, but, in the sense in which we have been discussing it, they are useless for description. (Where differences in the standard deviations of the predictors are important for the interpretation of the data, a situation which is relatively infrequently in psychological research, the raw score regression equation may properly be interpreted. See Ezekiel and Fox, 1959.)

When we wish to interpret or describe the composite in terms of its correlations with the predictor variables, we must compute these component loadings, using equation (6.9). The resulting values appear in the last row of the table.

A comparison of the $\beta$ weights with the component loadings reveals that the two methods for describing the composite, although yielding similar results, are not in complete agreement. One prominent example of this discrepancy between the $\beta$ weights and the component loadings is the variable years of college. Its $\beta$ weight ($-.19$) is the second largest in absolute value; however, its loading of $-.07$ is quite small, and it would not be considered important for describing the composite. Here we have an instance of a variable that makes one of the larger contributions to the variance of the composite by means of its $\beta$ but is almost unrelated to the composite in terms of that variable's correlation with the composite. The strange state of affairs occurs because the predictor variable is almost unrelated to the criterion. Geometrically, we have a situation similar to that shown in Figure 6-2*b*. The composite is almost at right angles to the variable, but the projection (parallel to the other predictors) of the composite onto the variable is definitely not zero, because of the location of the other variables. A simplified picture of how this might occur is presented in Figure 6-9. Here, $C$ has a low negative correlation with $X_1$, but because of the position of $X_2$, its $\beta$ weight is substantially larger.

The question of which indices, the $\beta$ weights or the component loadings, are better to use to describe the composite is sometimes asked. The answer depends on what one means by description. If one wishes to describe the composite in terms of the contributions the predictor variables make to its variance, then the $\beta$ weights are appropriate. If, on the other hand, one wishes to describe the composite in terms of its correlations or

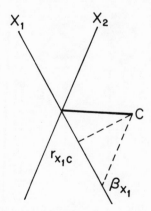

*Figure 6-9   Relationship between the component loading ($r_{X_1 C}$) and the regression coefficient ($\beta_{X_1}$). The loading is the projection of C perpendicularly onto $X_1$, while the regression coefficient is the projection of C onto $X_1$ parallel to $X_2$. The loading is the correlation between $X_1$ and C.*

relationships with the observed variables, then the component loadings should be used. Factor analysis generally involves description of the latter type because, as we have seen, a variable may either have substantial correlation with the composite while contributing relatively little to its variance (e.g., Home owned), or it may have little correlation with the composite and a relatively large $\beta$ weight (e.g., years of college). The decision as to which is the appropriate index depends on the investigators research objectives.

**Cross-validation.** The $\beta$ weights from Table 6-3 were used to compute predicted scores for the 30 individuals in the holdout group by multiplying the $\beta$s by the appropriate standard scores. The correlation between the resulting predicted scores and the actual salary scores of these 30 people is the cross-validation multiple correlation. The value of this correlation was .24 which, using the $t$ test for the statistical significance of a small correlation based on a small sample, does not reach conventional levels

**TABLE 6-4**

*Cross-Validation Component Loadings for the Regression Weights in Table 6-3.*

| Variable | Loading |
| --- | --- |
| 1 | .11 |
| 2 | .66 |
| 3 | −.18 |
| 4 | .14 |
| 5 | .07 |
| 6 | .09 |
| 7 | .14 |
| 8 | −.36 |
| 9 | .19 |
| 10 | −.29 |
| 11 | .35 |
| 12 | −.12 |
| 13 | .01 |
| 14 | −.27 |
| 15 | −.09 |
| 16 | .58 |
| 17 | .17 |
| 18 | .39 |

of significance ($p = .18$). It would be unwise to conclude that our $\beta$ weights will yield very accurate predictions for new samples.

On the other hand, the cross-validation component loadings in Table 6-4 show some similarity to the component loadings found in the development group. This is particularly true for the correlation of the composite with age and number of organizations. We might conclude from this that a component similar to the one found in the development group, but with less relationship with the time at present residence variable, exists in the cross-validation group. However, in this second group the composite does not appear to have an appreciable relationship with salary.

# 7

# CANONICAL ANALYSIS

In the last several years a number of authors (e.g., Dunnette, 1963) have criticized applied prediction studies because they felt that the use of a single criterion variable was a serious oversimplification of the problem. Years earlier, Hotelling (1935, 1936) provided a procedure that could handle the problem of complex criteria, but the method was so laborious to perform that it went totally unused except for a few demonstration problems. With the advent of large, fast computers and publication of the necessary programs (e.g., Cooley and Lohnes, 1962, 1971) a few applications of the procedure have appeared in the psychological research literature. The procedures described in this chapter, which Hotelling called *canonical analysis*, provide the most general case of external factor analysis. The other procedures described in this part (with the exception of Chapter 5) are all special cases of canonical analysis.

## GENERAL DESCRIPTION

Canonical analysis is a technique for finding the *correlations* between one set of variables, taken as a set, and a second set of variables, also

**175**

taken as a set. This may be contrasted with multiple correlation analysis, in which the *correlation* of a set of variables, taken as a set, is found with a single external variable. In fact, a multiple correlation analysis is a canonical analysis with only one variable in one of the sets. We might also say that a canonical analysis is a multiple correlation analysis when more than one criterion variable is being considered *simultaneously*.

Let us look at the geometric representation of canonical analysis to see how it relates to multiple correlation before we consider the method in detail. For an analysis to require canonical procedures, there must be two or more variables in each of two (or more!) sets. Four linearly independent, but correlated variables (the minimum number) require a four-dimensional representation, so we will have to rely on words primarily. Very briefly and simply, what a canonical analysis does is find the composite in each set of variables that has the maximum correlation with a composite in the other set. Each of these composites is defined by a set of regression weights in the same way that the composite of multiple correlation analysis is defined by its $\beta$ weights, and each is described by a set of component loadings that, once again, are the correlations of the observed variables with the composite. Actually, there is more than one composite in each set, but we confine our attention to only one for the present.

In multiple correlation analysis there is one variable in one of the sets and the other set contains $m$ variables. If we call the larger (predictor) set "set 1" and the smaller set "set 2" the composite of set 2 is the single variable in the set. There can be only one composite in set 2, and it must be the same as the observed variable because the one observed variable completely defines the one-dimensional set 2 space. Of the infinite number of composites of set 1 that could be found we select the one that has maximum correlation with the composite (the single variable) of set 2. The problem is relatively simple because there is only one possible composite in set 2.

When there is more than one variable in set 2, the problem calls for a canonical analysis. Here there are infinitely many composites in set 2 as well as in set 1, even when there are only two variables in each set. The task is to find the composite in each set that has maximum correlation with the composite of the other set. This requires that each composite be determined with reference to the other. Because at the outset neither of these composites is known, the solution to the problem might seem hopeless because we are trying to locate one unknown composite with reference to another, also unknown composite. The mathematical solution is extremely complex and almost unbelievably time consuming if done by hand, which explains why essentially no canonical analyses were done prior to 1960. However, presently available computer programs such as the one given by Cooley and Lohnes (1971) provide a canonical solution with relatively modest expenditures of computer time.

**The canonical composites.** It was mentioned above that there are really more canonical composites than one for each set. If we have two variables in each set, then the two variables in set 1 exist in a 2-space or plane, as do the variables in set 2. The first canonical composite in set 1 defines one dimension of the set 1 space, but there is an additional dimension of the plane that is not yet defined. Likewise, the first canonical composite of set 2 defines one of the two dimensions of that space. In each set there is an additional dimension of variation that is not defined by the first composite in the set. In general, a complete canonical analysis provides a composite for each dimension of the space of the smaller set of variables and a companion composite for each of these in the larger set.

Recall from Chapter 4 that when there are $n$ linearly independent variables, there are also $n$ dimensions of the space which contains these variables. In a canonical analysis we have two separate spaces. If there are $m$ variables in the larger set and $p$ variables in the smaller set, then the smaller set exists in a $p$-dimensional space. It is possible to find $p$-independent (uncorrelated) composites in this space because there are $p$-independent sources of variation. Canonical analysis provides a means of

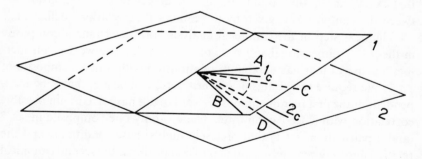

*Figure 7-1  Schematic vector representation of a canonical correlation. $1_c$ is a composite of A and B. $2_c$ is a composite of C and D. The canonical correlation is the correlation between $1_c$ and $2_c$.*

determining a particular set of $p$-dimensions in the smaller set and another particular set of $p$ dimensions ($p < m$) in the larger set.

Figure 7-1 is designed to give the reader a rough picture of what is going on in a canonical analysis. The figure is not at all accurate, because four dimensions are necessary. However, let us assume that plane 1, containing the set-1 variables $A$ and $B$, and plane 2, containing the set-2 variables $C$ and $D$, really do represent a 4-space. A canonical analysis might reveal a composite of $A$ and $B$ (called $1_c$ in the figure) in set 1 which has maximum correlation with the composite ($2_c$) of $C$ and $D$ in set 2. If the relationship between $1_c$ and $2_c$ were the highest that could be found for these data, these composites would be the first canonical composites of their respective sets. It is then mathematically possible (but not possible in our figure) to determine a second composite in each set that is independent of (uncorrelated with and, geometrically, at right angles to) the first pair of composites.

We now have a fairly complex situation. Let us assume that we have five predictor or set-1 variables and three criterion or set-2 variables. In this case a canonical analysis will first find *the* composite of the set-2 variables that has the highest possible correlation with a composite of set 1, and, simultaneously, the composite of set 1 that is most highly related to this set-2 composite will be determined. Remember that in the beginning there is an infinite number of set-2 composites from which we could choose, and, likewise, an infinite number of set-1 composites with which to pair it. We select our pair of composites subject to the restriction that this pair have the highest possible relationship. There is only one pair that exactly meets this condition, and the degree of the relationship between the composites is given by the canonical correlation coefficient.

The next step in the canonical analysis is to determine a composite in the $p - 1$ dimensional subspace of set 2 (in the case we are considering, a 2-space) that has maximum relationship with a new composite of set 1. We require two things of our new set-2 composite, that it be independent of the first pair of composites, and that it have maximum possible correlation with a set-1 composite. There is only one composite in set 2 (and a pairmate in set 1) that precisely fulfills these conditions, and the relationship between this second pair of composites is the second canonical correlation.

In our example the third composite of set 2 is now defined, because there is only one independent dimension left in the 3-space when the first two dimensions have been identified. (If there had been more variables we could have continued defining new composites, subject to the two restrictions mentioned above, until all dimensions of the smaller space had been exhausted.) Even though the last composite in set 2 is restricted by the earlier ones, we still must select the most highly related composite from set 1, because we still have three dimensions available in that set. The last and smallest canonical correlation is the index of relationship between these two final composites.

**Canonical correlations.** A canonical correlation may be viewed in a manner very similar to that of multiple correlation, although, in practice, it is computed rather differently. First, the canonical correlation ($R_c$) may be seen as a function of the angle between the two composite vectors. That is, if the variables in the two sets are plotted in a space of people and the subspace of variables has been defined, as has been done in Figure 7-1, then the composites are also in that space and can be represented by vectors. The cosine of the angle between the two composite vectors is the canonical correlation. In this sense it is just like the multiple $R$.

We can calculate scores for individuals on each of the composites by multiplying each person's standard scores for the variables in the set by the standard regression weights that define the composite. The simple product-moment correlation between the composite scores for set 1 and the composite scores for set 2 is the canonical correlation. Once again, the similarity with multiple correlation is obvious. The difference is that we must compute two sets of composite scores.

There is an important distinction that must be made between a composite and a variable when interpreting these correlations. When we compute a bivariate correlation, we have the simplest possible canonical analysis, because each set contains only one variable. The single variables take the place of the composites, and the cosine of the angle between the variable vectors is the correlation. In this case each composite accounts for all the variance of the single variable that makes up the set. We can, therefore, talk about the correlation (squared) as an index of the proportion of variance in one variable that is related to the other variable.

When we move to multiple correlation the situation changes for one of the sets, but, because we are generally interested in predicting a criterion set that is a single variable, this change goes unnoticed. It is legitimate to say that $R^2$ is the proportion of variance in the criterion variable that is related to the set of predictor variables. It is *not* legitimate to claim that

$R^2$ is the proportion of variance of the *predictor set* that is related to the criterion. What we can say is that $R^2$ is the proportion of the variance of the *predictor composite* that is related to the criterion variable.

The situation with canonical analysis is similar, except that now we have a set of criteria and a set of predictors. Because any single composite of either set cannot account for all the variance in the set, a correlation between a pair of composites can only indicate *the proportion of variance in each composite that is related to the other composite* of the pair. Thus the square of the largest canonical correlation is the proportion of variance of the first composite in one set that was accounted for by the first composite of the other set. Notice that $R_c^2$ is a symmetric index in the sense that the proportion of variance accounted for in the composite of one set is equal to the proportion of variance accounted for in the composite of the other (just as $r_{xy}^2$ is symmetric). This will not be true when we consider the problem of proportions of variance of the sets, which is discussed shortly.

**Weights and loadings.**   Canonical analysis can be viewed in two ways, either as a kind of super multiple correlation analysis or as an external factor analysis. In the first sense we attempt to predict a composite of one set of variables from a composite of the other. Each composite is defined by a set of regression weights which indicate the *independent* contribution of each of the variables to the composite. These weights can be used to calculate scores for each individual on the composites. Thus the canonical weights are directly analogous to the $\beta$ weights of multiple correlation analysis. Remembering our discussion of cross-validation in the last chapter, it is obvious that cross-validation is at least as important in canonical prediction as it is in multiple prediction. It might even be argued that cross-validation is more important for canonical analysis because there are two sets of weights, each of which will make maximum use of sample-specific covariation, rather than just one.

After the canonical weights have been obtained, it is possible to calculate the correlations of the variables in each set with the composites of the set. The correlations between the composites and the variables are called *canonical component loadings*. They may be calculated either by correlating scores on the composites with scores on the variables or by using the equations

$$\mathbf{S}_1 = \mathbf{R}_{11} \cdot \mathbf{B}_1$$
$$\mathbf{S}_2 = \mathbf{R}_{22} \cdot \mathbf{B}_2$$

$$(7.1)$$

In this formula, $S$ is the "structure" matrix of component loadings, $R$ is the matrix of correlations among the variables, and $B$ is the matrix of canonical weights. In $S$ the rows are variables 1 through $m$ (or $p$), and the columns are the first through $p$th composites; $B$ also has $m$ (or $p$) rows and $p$ columns. Note that this equation refers to only one set of variables at a time. A different $R$ and $B$ must be used to compute an $S$ for the second set.

**Mathematics.** The computations of a canonical analysis are more complex than any others we discuss in this book. For most problems of research interest, a computer is absolutely essential, as is a good computer program, such as the one given by Cooley and Lohnes (1971). Although the details of the mathematics are beyond our scope, a brief discussion of the elements of the solution may help the reader to appreciate more fully the numerical results of a canonical analysis.

Suppose that we have two sets of variables; set 1 has $m$ variables, and set 2 has $p$ variables. Let $R$ be the correlation matrix of all $m + p$ variables. Then we can break $R$ down into

$$
R = \begin{matrix} 1 \\ \vdots \\ m \\ m+1 \\ \vdots \\ m+p \end{matrix}
\begin{array}{c} \overset{1 \cdots m \mid m+1 \cdots m+p}{} \\ \left[ \begin{array}{c|c} R_{11} & R_{12} \\ \hline R_{21} & R_{22} \end{array} \right] \end{array}
$$

$R_{11}$ is the $m \times m$ matrix of correlations among the set-1 variables.
$R_{22}$ is the $p \times p$ matrix of correlations among the set-2 variables.
$R_{12} = R'_{21}$ are $m \times p$ and $p \times m$ matrices of correlations between the set-1 variables and the set-2 variables.

When there are more variables in one set than the other, the larger set is defined as set 1 (that is, $m \geq p$). This is done because, although the results are the same regardless of which set is called set 1, the computations are more efficiently performed when the larger set is called set 1.

When the data are set up in this manner, the solution of a set of matrix equations of the form

$$
(R_{22}^{-1} R_{21} R_{11}^{-1} R_{12} - \lambda_i I) b_{2i} = 0 \tag{7.2}
$$

under a standardizing restriction that

$$\mathbf{b}'_{2i}\mathbf{R}_{22}\mathbf{b}_{2i} = 1$$

gives us the $i$th squared canonical correlation as $\lambda_i$ and the vector of canonical weights in the smaller set (set 2) is contained in $\mathbf{b}_{2i}$. The vector of weights for set 1 ($\mathbf{b}_{1i}$) is computed from these results by

$$\mathbf{b}_{1i} = \frac{(\mathbf{R}_{11}^{-1}\mathbf{R}_{12}\mathbf{b}_{2i})}{\sqrt{\lambda_i}} \tag{7.3}$$

An equation of the form of (7.2) is called an eigenequation in which $\lambda_i$ and $\mathbf{b}_{2i}$ are unknown. There are many possible solutions for $\lambda_i$ and $\mathbf{b}_{2i}$ that would satisfy the equation, but only one solution will satisfy the restrictive conditions that the weights be standard regression weights and the relationship between the composites (the value of $\lambda_i$) be a maximum. The vector of weights is called an eigenvector, and $\lambda_i$ is known as an eigenvalue.

We noted above that there are $p$ independent composites in set 2 which might be found, each with a composite in set 1 with which it shows maximum relationship. Equation (7.2) gives the solution for any given one of these composites and its associated canonical correlation. By using (7.3), the weights for the composite in set 1 are determined. Because there are $p$ pairs of composites, there are also $p$ canonical correlations and $p$ pairs of vectors of weights. Thus the matrices $\mathbf{B}_1$ and $\mathbf{B}_2$ each have $p$ columns that contain the first through the $p$th sets of regression weights as determined sequentially by equations (7.2) and (7.3). There are $m$ rows in $\mathbf{B}_1$ corresponding to the $m$ variables in set 1 and $p$ rows in $\mathbf{B}_2$ for the $p$ variables in set 2. The $\lambda_i$ ($i = 1$ to $p$) are the squared correlations between the first through $p$th pairs of composites defined by these vectors of weights.

One is seldom interested in a complete canonical solution. Rather, attention is likely to focus on the larger or more salient canonical correlations and their associated composites. If the $\lambda_i$ are taken in order of their magnitude with the smallest taken first, a procedure developed by Bartlett (1941, 1947) may be used to test the statistical significance of each canonical correlation in turn. The necessary statistic is Wilks' lambda,

$$\Lambda = \prod_{i=1}^{p} (1 - \lambda_i) \tag{7.4}$$

in which it is important to remember that $\lambda_p$ is the largest squared canonical correlation. The test of the null hypothesis that all the $\lambda$s are no greater than would be expected by chance between unrelated sets of variables is given by

$$\chi^2 = -[(N - 1) - .5(m + p + 1)] \log_e \Lambda \tag{7.5}$$

in which $\chi^2$ has $m \times p$ degrees of freedom. A statistically significant value of $\chi^2$ is typically interpreted to mean that the largest canonical correlation is not zero. To test each succeeding $\lambda_i$, $\Lambda$ is recomputed as

$$\Lambda_k = \prod_{i=1}^{p-k} (1 - \lambda_i) \tag{7.6}$$

in which case, the test for the $k$th largest canonical correlation is

$$\chi^2 = -[(N - 1) - .5(m + p + 1)] \log_e \Lambda_k \tag{7.7}$$

with degrees of freedom $(m - k)(p - k)$.

It must be remembered, of course, that statistical significance is no guarantee that the relationship has any generality beyond the sample at hand. Nor is it an assurance that the results are useful scientifically. Statistically significant results can occur when the relationship is due solely to sample-specific covariation, in which case a cross-validation will point out the true source of relationship. Analyses using very large samples and relatively few variables may result in small correlations that are statistically significant but scientifically trivial. For example, with a sample of 8000 a bivariate pmc of .02 is statistically significant and may be highly reproducable, but it would be hard to claim that such correlations are of scientific importance. The investigator must decide for himself what level of correlation he regards as of scientific value, but it would seem reasonable in most cases to reject as meaningless a relationship in which $R_c^2 (= \lambda)$ is less than .10 (i.e., one composite accounts for less than 10% of the variance of the other).

**Sample size.** A review of the equations for testing for statistical significance of either the multiple [equation (6.15)] or canonical correlations [equations (7.5) and (7.7)] shows that sample size plays an important role here, just as it does in testing mean differences and bivariate correlations. However, the complex relations involved in multivariate analysis and the opportunity to capitalize on sample specific covariation make attention to sample size even more necessary in multivariate problems. From a purely practical standpoint, stability of results and success in cross-validation are much more likely if large samples are used (see the discussion of shrinkage in Chapter 6). Also, the investigator's claims to have discovered the "true" relationships among his variables are likely to be less in error when they are based on large samples.

The question of what constitutes sufficient sample size for a multivariate analysis does not, of course, have a simple and direct answer. Barcikowski and Stevens (1975) studied the stability of canonical weights and canonical loadings for varying sample sizes and found, as one would expect, that the sample size needed for stable results increases as the number of variables increases. Studies of the effect of sample size on the stability of factor loadings in internal factor analysis have come to the same conclusion.

One answer to the sufficient sample size question is "as many as you can get." However, two rules of thumb seem to offer somewhat more promise. Sample size should be related to the number of variables and should increase as the number of variables increases. One informal guide (and perhaps lower limit) is that there should be 10 subjects for each variable. We should probably add 50 to this number to ensure sufficient sample size for small sets of variables. Thus our first rule of thumb is that $N \geq 10 (p + c) + 50$.

A second and more stringent rule which is also a function of the number of variables is that $N$ should be equal to the square of the number of variables (and we should add 50 or 100 for small sets). This rule [$N = (p + c)^2 + 50$] means that our required sample size would rise rapidly as the sets become large. It gets logical support from the fact that the number of correlations among the variables rises much more rapidly than the number of variables does. Empirical support comes from the fact that the sample sizes recommended by Barcikowski and Stevens are surprisingly close to those provided by this second rule.

In a practical situation with a fairly large number of variables it is quite unlikely that the investigator will have access to the number of subjects called for by the second rule (1650 for a 40-variable problem). Under these conditions he should attempt to satisfy the first rule (450) and, failing that, he should proceed to analyze the data of as many subjects as possible *while holding out a group for cross-validation*, and he should exercise extreme caution in his interpretations.

**Redundancy.** As we noted above, the meaning of a canonical correlation can be misinterpreted by the unwary (and there are examples in the literature where this has happened) because it refers to the relationship between a pair of composites of the two sets, not to the relationship between sets themselves. To see that this is true we need only consider the canonical component loadings, the correlations between the variables in a set and a composite of the set. Each of these loadings is a bivariate correlation, the square of which can be interpreted as the amount of variance of the variable that is accounted for by the composite. By taking the sum of squared loadings for a particular composite, we get the amount of variance of the set that is accounted for by the composite. Because each variable has unit variance, we can divide the sum of squared loadings by the number of variables in the set (which is the total variance of the set) to obtain the proportion of the variance of the set that is accounted for by the composite. If we now multiply this value by the squared canonical correlation, the result is the proportion of variance in one set that is accounted for by the composite of the other set. Doing this for all the composites of a set and summing the results yields the proportion of variance of one set that is accounted for by the other set.

The procedure just described was developed by Stewart and Love (1968), who gave the name *redundancy* $(R_d)$ to the resulting indices. There is a redundancy index for each composite in each set which is given by the general equation

$$R_{d_{1i \cdot 2}} = \left[ \frac{\sum\limits_{k=1}^{m} S_{1ik}^2}{m} \right] R_{c_i}^2 \tag{7.8}$$

for the $i$th composite of set 1. In this equation $S_{1ik}$ is the correlation of the $k$th variable in set 1 with the $i$th composite of set 1, and the redundancy index $R_{d_{1i \cdot 2}}$ is the proportion of variance in set 1 that is accounted for by the $i$th composite of set 2. Similarly,

$$R_{d_{2i \cdot 1}} = \left[ \frac{\sum\limits_{k=1}^{p} S_{2ik}^2}{p} \right] R_{c_i}^2$$

yields the proportion of variance of set 2 that is accounted for by the $i$th composite of set 1. Each of these $R_d$s is a redundancy index. The total redundancy of set 1 with set 2 (the proportion of the set 1 variance that is related to set 2) is given by

$$\bar{R}_{d_{1 \cdot 2}} = \sum_{i=1}^{p} R_{d_{1i \cdot 2}} \tag{7.9}$$

A similar equation gives the redundancy of set 2 with set 1.

It is important for the reader to realize that redundancy is not necessarily symmetrical. In the case of a bivariate relationship the interpretation of the squared correlation is symmetrical in that it indicates the proportion of variance in each variable that is related to the other. When we are dealing with either multiple or canonical correlations the symmetry refers only to the composites. To see that symmetry does not generalize to the sets themselves, suppose that set 1 is composed of variables that are highly correlated with each other and set 2 is made up of variables that have essentially zero correlations. Suppose also that one of the variables in set 2 is highly correlated with the variables in set 1. The other set-2 variables must then have low correlations with the set-1 variables. In cases such as this the first canonical component of set 2 will be highly related to the one variable that has high correlations with set 1 and will account for approximately $1/p$ of the set-2 variance. The set-1 component, on the other hand, will be highly related to most of the variables in set 1 and will account for a relatively large proportion of the set-1 variance. This will result in a high redundancy of set 1 with set 2 and a low redundancy of set 2 with set 1.

Another reason why we would not expect redundancy to be symmetrical is that the two sets may have very different numbers of variables. The mathematics of canonical analysis are such that the $p$ components of set 2 that are obtained from a complete canonical analysis will account for all the variance in set 2. However, the $p$ components that are extracted from the $m$ variables of set 1 cannot account for all the set-1 variance. If they did, there would be linear dependence in set 1, $\mathbf{R}_{11}$ would be singular, and it would be impossible to obtain a solution using equation (7.2) because $\mathbf{R}_{11}^{-1}$ would not exist. Therefore, the total redundancy of the smaller set with the larger set will always be greater than the total redundancy of the larger set with the smaller.

There is another aspect of redundancy that points up nicely the major difference between internal and external factor analysis. When an internal factor analysis is performed the first composite is determined in such a way that the value of $\Sigma s^2$, the sum of squared component (or factor) loadings is maximized, and each succeeding composite accounts for the maximum amount of the remaining variance (see Chapter 10). It is quite often the case in a canonical analysis that some composite other than the first will have the highest redundancy, because the pairs of composites are considered in order of their correlations with each other, not in order of the proportions of variance they account for in their respective sets. This phenomenon is most likely to occur when there is a single variable in set 1 that is highly correlated with a single variable in set 2. When this happens it may be desirable to ignore the first pair of composites and consider only the later ones. However, it may also be argued that these variables should not have been included in the analysis in the first place, in which case the analysis should be rerun with the offending variables omitted. The computer allows us the luxury of this type of error at relatively little cost in time.

At the risk of being redundant on the matter of redundancy, we will take a look at the issue from another perspective. We have seen that it is possible to calculate the correlations of the set-1 variables with the set-1 composites. Well, it is also possible to calculate the correlations of the set-1 variables with the set-2 composites. The same may be done for the set-2 variables and set-1 composites. This may be done either by computing the composite scores and then the correlations or by using the equations

$$\mathbf{S}_{1_{c2_i}} = \mathbf{R}_{12}\mathbf{b}_{2_i} = \mathbf{S}_{1_i}\mathbf{R}_{c_i}$$
$$\mathbf{S}_{2_{c1_i}} = \mathbf{R}_{12}\mathbf{b}_{1_i} = \mathbf{S}_{2_i}\mathbf{R}_{c_i} \tag{7.10}$$

That is, the correlations of the set-1 variables with the $i$th set-2 composite are obtained by multiplying the interset correlation matrix $\mathbf{R}_{12}$ by the $i$th vector of set-2 regression weights (or by multiplying the $i$th vector of set-1 component loadings by the $i$th canonical correlation), and the correlations of the set-2 variables with the set-1 composites are obtained in a similar manner. In either case, the means of the sums of squared correlations (loadings) give the redundancy indices directly, because each of these squared correlations is a proportion of common variance between a variable in one set and a composite in the other.

**Stepwise canonical analysis.** It was stated in our discussion of multiple correlation that it is often desirable to find some reduced set of variables that will predict the criterion almost as well as the full set of predictors will. The chief gains to be realized from a stepwise approach, increased stability of the regression equation and greater simplicity of description, apply with at least as much force in canonical analysis.

We noted above that any least-squares procedure will capitalize on chance relationships that are due to sample-specific covariation. As the number of variables increases, the probable effect of these sources of variation on the canonical correlations increases. Therefore, the fewer variables there are in a canonical analysis which yields a correlation of a given magnitude, the greater is the likelihood that that correlation is due to real, population-wide sources of covariation, rather than to sample-specific sources. In an unpublished study Thorndike and Weiss (1969) found that elimination of a fairly large proportion of the variables in either set does not have a substantial effect on the magnitude of the first canonical correlation. The same study, using a double cross-validation procedure, showed that the canonical regression equations were more stable when irrelevant variance was removed by dropping those variables that did not contribute to the equation.

There is another aspect of the stepwise procedure applied to canonical analysis that has not received the attention it deserves. It may be possible to improve the content of a battery of predictors that are to be used for predicting a complex criterion by using an empirical step-up or step-down approach. Let us see how it might apply.

There is ample empirical and theoretical justification to support a statement that a single criterion measure is seldom a very accurate indicator of the success of a selection procedure in an industrial or other personnel setting (e.g., Dunnette, 1963). One is often interested in such diverse aspects of job performance as quality of output, quantity of output, absenteeism, tenure on the job, and relationships with co-workers. Similarly, there are many outcome variables that are of interest to a psychotherapist. Traditionally, when an investigator has tried to predict a combination of criterion variables, the approach has been to form a single composite of the several variables, based on their subjective importance and use the multiple regression equation that best predicts this single complex criterion. The variables may or may not be differentially weighted. Thus one industrial psychologist might consider quality and quantity of output, absenteeism, tenure, and co-worker relations all to be of equal importance, in which case he would obtain his composite criterion scores by simply summing each person's (hopefully standard) scores on the five variables. Another investigator might feel that the output variables are twice as important in the definition of the criterion as the other variables. For him, the composite would be (2 × quality) + (2 × quantity) + absenteeism + tenure + relationships, again using standard scores. In either case, a multiple correlation analysis using the predictor variables to predict the a priori weighted composite would be used. (Note that it is necessary to use standard scores in either situation; otherwise the variables will receive weights proportional to the variability of scores on the variables. Unless some adjustment is made for differing variabilities the resulting composite will not have the desired composition. This, of course, is related to the difference between $\beta$ weights and raw score regression coefficients discussed in Chapter 6).

A situation such as the one described above is made to order for a canonical analysis. However, when canonical analysis is suggested and explained to investigators with problems such as these, the immediate and almost universal response is something like this, "Well, that's fine, but what if the criterion composite weights don't come out as I think they should?" This is a valid criticism. The mathematics of canonical analysis pays no attention to the value of the variables for anything but least-squares prediction. The weights will be determined to yield the criterion that can be most accurately predicted, and this may mean that some important criterion variables will receive negligible weights.

One solution to problems such as this is to use an empirical stepwise procedure that involves a priori judgments about the relative importance of the several criteria and trying out several combinations of predictors to determine the set that results in the best prediction of a satisfactory composite. After the investigator has decided on the relative importance each criterion variable should have in the final criterion composite, he should perform a complete canonical analysis using these criterion variables and the predictor variables that, in his judgment, should give the best results. The canonical analysis will tell him what his selection of predictor variables can do. If one predictor dominates the analysis because of a high correlation with one of the criterion variables, this predictor may be eliminated and a new analysis performed. By selectively dropping predictors, it may be possible to identify a subset of the original variables that has substantial correlation with the desired criterion composite.

Canonical analysis may also be used in the manner described above for battery development. It may not be possible to find a set of predictors from among those available that will yield the desired criterion composite. This signals the need to add new variables to the predictor set, variables that have high correlations with those criterion variables which are not receiving large enough weights. In this way, an improved battery of predictors may be developed that will do the job most efficiently, because areas in which prediction is weak can be identified. The understanding of what is predicting what, which is necessary for this type of battery improvement, cannot be gotten from the more conventional multiple correlation analysis using a priori weights.

Of course, it may not be possible to find any set of predictors that will adequately predict a single composite of the criteria and will yield the desired relative weights. This is most likely to happen when the correlations among the criteria are low, making it difficult or impossible to generate a composite with the desired properties. In such a case the canonical analysis may still give useful results through one of two courses of action. Because the criterion composites are independent of each other, some combination of composites may prove useful. That is, criterion composite scores might be summed to yield the desired relative weights and the predictor composite scores summed to give the best prediction. The author knows of no situation where this has been attempted, but the idea is inviting.

If all else fails, the investigator can still fall back on the traditional approach of defining the composite using a priori weights. The exercise in canonical analysis will at least have provided a more thorough knowledge of the internal structure of his problem. Naturally, the results of any analysis described in this section must be subjected to a cross-validation before confidence can be placed in the generalizability of the results. (See Thorndike [1977] for a more complete discussion and an example).

**Complex composite prediction.** Canonical analysis provides another approach to the prediction problem which has not seen empirical application at this level of complexity. There is a selection strategy which has been used with pairs of predictor variables called *conjoint prediction*. Given two variables, both of which are related to the criterion, it is possible to choose a score on each of the variables so as to maximize the probability of success of the individuals selected. If we have two predictor variables, each of which is positively correlated with the criterion, then an individual who is high on one of the predictors is more likely to be high on the criterion than one who is low on that predictor. By selecting only those individuals who are high on both of the predictors, we increase still further the probability that the people we select will be successful. (Notice that this strategy is quite different from that which is involved in multiple correlation. An individual who is extremely high on one variable and average or low on the other could win a judgment of "predicted success" when multiple regression procedures are used. Conjoint prediction is more conservative in that it requires high scores on both of the predictor variables.)

The conjoint prediction strategy can be applied quite nicely to canonical prediction problems. Suppose, for example, that the first two canonical correlations prove to be statistically significant. This means that there are two independent sources of variation in the criterion set that are related to the predictor set. If satisfactory performance can be assumed to be dependent to some extent on high performance on all the criterion variables, then the most satisfactory individuals should be those who score high on both of the criterion composites. Because a high score on each of the first two predictor composites is indicative of a (probable) high score on the related criterion composite, those individuals who score high on both of the predictor composites show the greatest promise of being satisfactory. The fact that the second composite contains information that is independent of the first means that this strategy will make the most efficient use of the information available.

This procedure can also be used to solve the problem of multiple criteria when multiple regression is used for prediction. A separate regression equation would be determined for each of the criteria, and only those individuals with high predicted scores for all criteria would be selected. It would even be possible to weight differently the predicted criterion scores to reflect their relative subjective importance (e.g., by setting higher cutting scores for the more important criteria). However, from a purely statistical point of view this approach would not be as efficient as the canonical procedure, because the criterion variables would not be uncorrelated. To the author's knowledge neither of these procedures has ever been put to empirical test; thus, their relative merits for a practical problem are unknown.

**Cross-validation.**   The possible effects of sample-specific covariation on the regression equations that result from a canonical analysis are substantially greater than is the case for multiple regression equations. This is so because sample-specific sources of covariation may be present among the criteria as well as among the predictors. In addition, there are $p \times m$ relationships between criteria and predictors that can be subject to these chance effects rather than just the $m$ relations present in multiple regression. It is, therefore, even more imperative that canonical prediction equations be cross-validated than it is for multiple regression equations.

It is relatively simple matter to cross-validate a canonical equation by either the holdout group or double cross-validation method. If, as is almost always the case, a correlation matrix has been analyzed,* the canonical weights should be applied to standard scores. The predictor and criterion standard scores of each individual in the cross-validation group are multiplied by the appropriate canonical weights for a pair of composites and summed to yield scores on the two composites. The product moment correlation between these composite scores is the cross-validation canonical correlation. Although this correlation may be tested for statistical significance by the procedures appropriate for bivariate correlations, its magnitude relative to the original canonical correlation is generally of more interest than the probability that it is zero in the population.

In practice, canonical analysis has seldom been used for prediction. It is used much more frequently for external factor analysis to answer questions about related dimensions in two sets of variables. In such situations interest centers around the canonical component loadings that describe the composites as well as around the canonical correlations. Discussions of results are couched in terms similar to those of traditional factor analysis; for example, "component $i$ in set 1, which may be interpreted as such and such because of its high loadings on variables $X$, $Y$, and $Z$, was found to have a correlation of $R_c$ with a component of set 2 interpreted as thus and so (high loadings on $A$ and $B$, negative loading on $C$)." Although we might consider this type of interpretation justified because of prevailing practice in internal factor analysis, we may reasonably ask whether cross-validation has a proper place in this type of research.

A partial answer to the question of the place of cross-validation in external factor analysis has been provided by Thorndike and Weiss

---

* It is also possible to analyze variance-covariance matrices or cross-product matrices, but this is seldom done. The necessary equations are given by Morrison (1967).

(1973). In a study utilizing two large samples, each split into random halves for double cross-validation and with different sets of variables in the different samples, these authors found that the issue is not clear. On one hand, they found that the *canonical weights* may not be stable. Pairs of composites that were significantly correlated in the group on which they were developed were, in some cases, uncorrelated in the cross-validation group. On the other hand, the canonical components were found to be highly stable, meaning that the correlations between the variables and the composites did not change in going from the development to the cross-validation group. We may summarize these findings as meaning that the interpretations of the two canonical components in a pair will not change when the weights are applied to a new sample, but our conclusions about the degree of relationship between those components may require drastic modification. In other words, results from our development sample may indicate that two components, $X$ and $Y$, are correlated, but when the weights are applied to a new sample, we may be forced to conclude that the same two components are independent.

An alternative to cross-validation in external factor analysis is replication. If the same data are collected from two samples and independent canonical analyses are performed, the results may be compared for similarity without the guarantee of essentially identical canonical components that follows from a cross-validation. (Obviously, a replication study is contained within a double cross-validation.) However, replication also has its problems, as is evident from the attention given to the factor-matching problem of traditional factor analysis (see Chapter 12). A component that appears in the first pair in one analysis may be second or third in order in a replication. If a pair of components is clearly identifiable in both samples, there is no problem; however, in this case it is also highly likely that a cross-validation would have shown the weights to be stable. It would appear that the best approach is to use a double cross-validation design so that both the stability of the weights and the generality of the components can be evaluated.

## AN EXAMPLE

For our example of canonical analysis we return to the data presented in Chapter 6. However, this time we shall consider the prediction of all four criterion variables that were available: level of occupation, tenure,

**TABLE 7-1**

*Correlations of 4 Occupational Variables with 18 Biographical Variables from Table 6-3 and Correlations among 4 Occupational Variables.*

| | 1 | 2 | 3 | 4 | 5 | 6 | 7 | 8 | 9 | 10 | 11 | 12 | 13 | 14 | 15 | 16 | 17 | 18 | A | B | C | D |
|---|---|---|---|---|---|---|---|---|---|---|---|---|---|---|---|---|---|---|---|---|---|---|
| A | −08 | −15 | −16 | −09 | 00 | −04 | 15 | −10 | 01 | 01 | −03 | 21 | 06 | 02 | 04 | −10 | −05 | 10 | 1.00 | | | |
| B | 01 | 64 | −12 | 22 | −07 | 03 | −14 | −08 | 22 | −21 | 33 | 01 | −07 | 01 | −13 | 33 | −07 | −06 | −15 | 1.00 | | |
| C | 09 | 38 | 00 | 21 | −01 | 10 | −04 | −04 | 12 | −24 | 38 | 00 | 08 | 11 | 01 | 35 | 13 | 10 | −09 | 47 | 1.00 | |
| D | −15 | 23 | −02 | 19 | 06 | 10 | −11 | −01 | 14 | −05 | 08 | 01 | −14 | −07 | 05 | 21 | 22 | 05 | −10 | 42 | 46 | 1.00 |

salary, and supervisory responsibility. We use the same 18 predictor variables, whose correlations are given in Table 6-3, and determine the optimum composites of the two sets. In addition to the correlations in Table 6-3, we need the correlations among the criteria and the correlations of the predictors with the criteria. These are given in Table 7-1.

The results of the complete canonical analysis are given in Table 7-2, where we find the canonical weights, canonical correlations, and significance test information, and in Table 7-3, which contains the canonical component loadings and redundancy indices. In each of these tables the information for all four possible pairs of composites is given.

These data illustrate several of the points made. For example, the composites in the first pair are dominated by a single bivariate relationship between age in set 1 and tenure in set 2. Their bivariate correlation is .64, and each has a very high weight for the first composite in its set. Only GPA and years of college made additional contributions to the composite in set 1, whereas salary adds some of its variance to the set-2 composite. The fact that the large weights have a negative sign need not affect our interpretation because we can change all the signs of the weights in both composites without affecting the relationship. (If we change any, we must change them all.) Thus we may say that the set-1 composite is composed primarily of variation in age, whereas the first set-2 composite reflects variation in length of service. The $\chi^2$ test indicates that the remaining composites are not statistically significant; thus they would not ordinarily be given further consideration.

Comparison of the first canonical component loadings (column 1 of Table 7-3) with the canonical weights (column 1 of Table 7-2) illustrates the need for deciding what type of interpretation is desired of a canonical analysis. The loadings would lead to a different interpretation than that given above for the weights. Although age retains the most prominent position, time at residence and number of organizations also have substantial correlations with the composite. Neither of these variables receives a large weight, but both are variables that should show increasing values with increasing age. Thus they reconfirm our suspicion that the predictor composite is a combination of age and its correlates. If we wished, we might label this composite maturity (remember that we may reverse all signs of the loadings and weights). Note that the variable with the second highest weight, years of college, has little correlation with the composite. Although this does not always happen, its low loading simplifies our interpretation of the composite.

**TABLE 7-2**

*Canonical Weights, Canonical Correlations and Tests of Significance
from the Canonical Analysis of 18 Biographical Variables and
4 Occupational Variables.*

| Predictor Variable | Predictor Composites | | | |
|---|---|---|---|---|
| | 1 | 2 | 3 | 4 |
| 1 | .02 | .07 | −.46 | .19 |
| 2 | −.92 | .03 | .09 | .66 |
| 3 | .03 | −.12 | −.50 | .10 |
| 4 | −.24 | −.17 | −.06 | .06 |
| 5 | −.06 | −.20 | .03 | −.16 |
| 6 | −.11 | −.29 | −.08 | −.10 |
| 7 | −.03 | .27 | .12 | −.32 |
| 8 | .32 | −.09 | .04 | .22 |
| 9 | −.15 | .15 | .26 | −.16 |
| 10 | .00 | .00 | .00 | .49 |
| 11 | .05 | .17 | −.31 | −.62 |
| 12 | −.06 | .04 | .48 | −.44 |
| 13 | .01 | .27 | −.42 | .15 |
| 14 | .02 | .54 | −.26 | .17 |
| 15 | .02 | −.56 | .04 | −.15 |
| 16 | −.10 | −.43 | −.03 | −.27 |
| 17 | −.01 | −.67 | −.22 | −.09 |
| 18 | −.15 | .26 | .12 | −.30 |

| Criterion Variable | Criterion Composites | | | |
|---|---|---|---|---|
| | 1 | 2 | 3 | 4 |
| A | −.01 | .39 | .61 | −.71 |
| B | −.88 | .49 | .41 | .46 |
| C | −.24 | .05 | −.89 | −.77 |
| D | .03 | −1.01 | .56 | −.12 |
| $R_c$ | .73 | .48 | .42 | .39 |
| $\Lambda$ | .25 | .54 | .70 | .85 |
| $\chi^2$ | 119.59 | 53.56 | 30.84 | 14.13 |
| df | 72 | 51 | 32 | 15 |

The nature of the criterion composite also changes when the loadings rather than the weights are used for interpretation. Tenure becomes even more prominent, but salary and supervisory responsibility gain in importance. Again, the composite seems strongly related to time. The pair of composites confirm a picture one might predict, given the nature

**TABLE 7-3**

*Component Loadings and Redundancy Indices for the Canonical Analysis.*

| Predictor Variables | 1 | 2 | 3 | 4 |
|---|---|---|---|---|
| 1 | −.04 | .26 | −.49 | .04 |
| 2 | −.89 | .08 | −.08 | .21 |
| 3 | .15 | −.21 | −.39 | .16 |
| 4 | −.33 | −.24 | −.10 | −.03 |
| 5 | .09 | −.20 | .04 | −.07 |
| 6 | −.06 | −.21 | −.10 | −.10 |
| 7 | .18 | .20 | .04 | −.33 |
| 8 | .10 | −.15 | −.15 | .18 |
| 9 | −.30 | −.05 | .17 | −.05 |
| 10 | .34 | −.13 | .26 | .23 |
| 11 | −.53 | .18 | −.43 | −.33 |
| 12 | .02 | .17 | .33 | −.37 |
| 13 | .05 | .27 | −.34 | −.31 |
| 14 | −.05 | .18 | −.29 | −.21 |
| 15 | .16 | −.21 | −.03 | −.27 |
| 16 | −.50 | −.16 | −.30 | −.21 |
| 17 | .05 | −.57 | −.12 | −.32 |
| 18 | .03 | −.07 | −.06 | −.47 |

| Criterion Variables | Criterion Components | | | |
|---|---|---|---|---|
| | 1 | 2 | 3 | 4 |
| A | .13 | .41 | .58 | −.69 |
| B | −.98 | .03 | .13 | .15 |
| C | −.64 | −.23 | −.50 | −.54 |
| D | −.45 | −.83 | .26 | −.22 |
| $R_{d_{1 \cdot 2}}$ | .051 | .011 | .011 | .009 |
| $R_{d_{2 \cdot 1}}$ | .210 | .052 | .029 | .031 |

of the variables. We might even be tempted to label the criterion composite "occupational maturity."

The redundancy indices illustrate the asymmetry discussed earlier and also give us little cause for joy concerning our results. With a canonical correlation of .73 we find only 21% of the variance in the criterion set is variance shared with the predictor set. Utilizing all four composites, we find only 31.2% of the criterion set variance to be redundant with the predictor set. Looking the other way is even worse. Only 8.2% of the total predictor variance is redundant with the criterion set, and most of this is from the first composite. This depressing state of affairs is due to the relatively low within set correlations found in $R_{11}$ and $R_{22}$.

The picture presented by the cross-validation of the canonical weights on our 30-member holdout group is quite positive, although before we

**TABLE 7-4**

*Cross-Validation Component Loadings and Canonical Correlation.*

| Variable | Predictor Component Loadings | Variable | Criterion Component Loadings |
|---|---|---|---|
| 1 | .07 | A | .48 |
| 2 | −.91 | B | −.98 |
| 3 | .38 | C | −.51 |
| 4 | −.22 | D | .04 |
| 5 | .04 | | |
| 6 | −.03 | | |
| 7 | −.05 | | |
| 8 | .29 | | |
| 9 | −.33 | | |
| 10 | .13 | | |
| 11 | −.13 | | |
| 12 | .09 | | |
| 13 | .08 | | |
| 14 | .10 | | |
| 15 | .14 | | |
| 16 | −.56 | | |
| 17 | .20 | | |
| 18 | −.23 | | |

Cross-validation
$R_c = .71.$

celebrate with champagne we should remember that it is primarily the relationship between age and tenure we are cross-validating. Table 7-4 contains the component loadings and canonical correlation that resulted from the application of the first set of weights to the standard scores of the 30 people in the cross-validation group. The canonical correlation, which is the bivariate correlation between composite scores, shrank very little, indicating that the relationship between composites defined by the obtained weights is a general characteristic of the population.

The cross-validation component loadings do not give clear confirmation of our interpretation of the composite. The high loadings for age and number of organizations are consistent with the results from the development group, but the loading for time at residence drops considerably. Several other loadings show modest changes of .10 or .20. On the criterion side, the loading for level of occupation rises, but the supervisory responsibility loading drops. The stability of these loadings is less than that found in previous research and may in part be because the age-tenure relationship dominates the composites. In general, the other weights are probably too small to indicate real and important sources of variation. Another factor that may have contributed to the instability of the loadings is the size of the two samples. Neither sample is really large enough to justify a canonical analysis with 22 variables. Our liberal rule of thumb $[N = 10(p + c) + 50]$ would call for 270. Of these, perhaps 200 should be used for development and 70 for cross-validation.

## SUMMARY

In this chapter we have discussed canonical analysis as a generalization of multiple correlation analysis. Because it is not realistic to undertake hand computation of a canonical analysis, our attention has focused on interpretation of results and on research strategies. We have seen that canonical analysis involves finding the composites in one set of variables that are maximally correlated with composites in a second set of variables. The composites are defined by canonical regression weights (which are generally in standard score form), and they can be described by the canonical component loadings. The canonical correlations are the pmc's between pairs of these composites, and the pairs of composites are ordered by the magnitude of their correlations. The largest canonical correlation will be at least as large as the largest multiple correlation of a single variable in either set with the entire other set.

We noted that the interpretation of a canonical correlation must be phrased in terms of the composites rather than the sets of variables. When it is desirable to discuss proportions of variance of the sets, the concept of redundancy must be used. The component loadings are used to compute the proportion of variance that a composite accounts for in its own set as well as its redundancy.

Stepwise procedures may be used beneficially in both predictive and descriptive applications of canonical analysis. When the objective of a study is prediction, it is absolutely necessary that some form of cross-validation be performed. The best research design for external factor analyses appears to be double cross-validation, because it provides both cross-validation and replication. The latter may help one evaluate the generality of the canonical components, whereas the former indicates the stability of the canonical weights.

## BEYOND CANONICAL ANALYSIS

By this time the reader probably has thoughts running through his mind such as, " Good grief, what could they possibly do to me beyond canonical analysis? What would be more complicated than that?" Well, there are two rather painless and potentially useful extensions of the ideas discussed thus far. One of these, which we might call multiple canonical correlation, was developed by Horst (1961) and involves finding the relationships among three or more sets of variables simultaneously. Although this procedure may be useful in some situations, we mention it only in passing because another procedure, which may be considered an outgrowth of it, is of greater potential research interest.

Cooley and Lohnes (1971) have described a statistical procedure which they call *multiple partial correlation* and have presented a program for its computation. In its simplest form, multiple partial correlation involves finding the correlations among the variables in one set after partialling out relationships with all variables in another set. Expressed in this way, we are dealing with a generalized partial correlation problem that has as its simplest case the partialling of one variable from the relationship between two others. When each set contains two variables, we have a second-order partial correlation problem such as the one described in Chapter 5. For problems involving more than two variables in each set it is computationally easier to work with matrices.

The problem of multiple partial correlation may be viewed in another way. Each matrix of correlations among the variables in a given set ($R_{11}$ and $R_{22}$) defines the relations of the variable vectors in a space with as many dimensions as there are variables in the set. The space for each set has some overlap with the space of the other set, which is given by the interset correlations ($R_{12}$). If we were to remove the effect of the set-2 variables from the relations among the set-1 variables we would be expressing the relations among the set-1 variables in a space that is independent of the set-2 space. This means that each variable vector in set 1 would be projected into that portion of the set-1 space that is independent of the set-2 space. These projections of the set-1 variable vectors are called residual vectors, and the space into which they are projected is the residual space. The relations among these residual vectors are the residual correlations or multiple partial correlations among the set-1 variables.

Although multiple partial correlation is of interest in its own right, it is perhaps of greater value when applied to problems of a canonical type. Consider, for example, a situation in which we have obtained several measures of academic potential and measures of interest in several academic areas. These might be such things as scores on the Differential Aptitude Tests (DAT) factors and the college major scales of the Kuder Occupational Interest Survey. It might be of interest to know how DAT scores relate to course grades when the effects of variations in interest are removed from the course grades. If we are interested in the simple correlations between scores on single tests and grades in single courses, our problem is one of multiple part correlation, for which equations are given by Cooley and Lohnes (1971, p. 203). By calculating the appropriate matrices of residual correlations, we may also compute a part (or partial) canonical analysis. This approach would give us a canonical analysis between the set of ability measures and the set of achievement measures cast in the residual space that is independent of the interest measures.

This section has been included to give the reader a few ideas about possible future trends in multivariate correlational analysis. The mathematics underlying these procedures has been developed, and some programs are available to perform the necessary computations. However, substantive studies using multiple canonical analysis or the higher-order partialling techniques outlined above are virtually nonexistent. It is, therefore, not possible to give any clear advice about the best methodological approaches to use. These procedures have the potential of great descriptive and perhaps explanatory power when their properties are better understood. They are presented here in the hope that knowledge of their existence by more investigators will lead to increased use. It may then be possible to determine what methodologies and types of interpretation are most justifiable.

# 8

# DISCRIMINANT ANALYSIS

There is a special application of multiple or canonical correlation analysis that can be very useful for certain research problems. Suppose we have collected data on several variables from two or more different groups of subjects and wish to describe the ways in which the groups differ on these variables. One way to attack the problem is to examine the group means and variances directly and describe the differences among the groups on each variable in turn. However, it may also be desirable to seek one or more composites of the variables such that the composites show maximum differences among group means of composite scores and minimum overlap in the distributions of these scores. An approach such as this is called a *discriminant function analysis*, because its objective is to find functions (composites) of the variables that maximally discriminate among groups. Depending on the number of groups, it is a special application of multiple regression or canonical analysis.

A second way to view this situation is as a prediction problem. We have seen in several places that we can predict an individual's score on a particular criterion variable by knowing his group membership or, in the case of a continuous predictor, his score on the predictor variable and the least-squares regression equation. Suppose we turn the situation around and now say that we wish to predict a person's group membership from his score on a continuous variable. The situation might be somewhat like that depicted in Figure 8-1, in which the distributions of scores for two groups on a variable, $X$, are shown. The dotted line marked "cutting score" ($X_c$) is the value of $X$ where the smallest number of incorrect decisions about group memberships would be made. If a person's score is greater than the cutting score, we would predict that he is a member of group 2, whereas we predict membership in group 1 if his score does not exceed the cutting score. Using a value of $X$ other than $X_c$ as the cutting score will result in a greater total percentage of errors of prediction than necessary.

At this stage we are not really talking about a discriminant function in the usual sense, because we have only one variable. However, we are making discriminations between groups as a function of their distributions of scores on the variable; thus we can speak of this as a discriminant function analysis in a simplistic sense. The power of the procedure does not become obvious until we add additional variables and use a composite of those variables to make our predictions.

We can carry the prediction analogy a little farther by noting the similarity between the situation in Figure 8-1 and the point biserial correlation. Group membership is a dichotomous variable; thus the ability of our measured continuous variable to predict group membership is reflected by $r_{pb}$ in the sense that lower overlap between the groups yields a higher correlation and more accurate prediction of group membership.

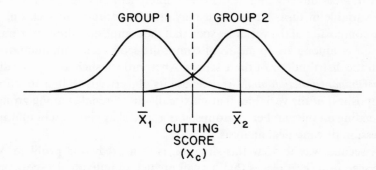

*Figure 8-1  Optimum cutting score to discriminate between two groups on the basis of one variable.*

Discriminant function analysis involves the generalization of this principle to more than one predictor variable and the possibility of more than two groups.

**Dichotomous criterion variables.** One of the important concepts to grasp in understanding how discriminant analysis works is the way in which group membership information is translated into dichotomous variables. When there are two groups the problem is simple. If we have groups $A$ and $B$, we can code individuals as either 1 or 0, as we did for a point biserial correlation. A code of 1 means membership in group $A$ and 0 means not membership in $A$. Because all our subjects are members of $A$ or $B$, not $A$ means membership in $B$, and a single dichotomous variable is all that is necessary to specify group membership completely.

Now, let us suppose that we have three groups of individuals, those who are members of group $A$, those in group $B$, and those in group $C$. Using one dichotomous variable, we can code those subjects who are members of group $A$ as 1 and those who are not members of $A$ as 0, but we need another variable to distinguish between groups $B$ and $C$. By specifying that a 1 on a second dichotomous variable means membership in group $B$ and 0 means not a member of $B$ we can completely specify group membership as follows:

$$10 = \text{member of } A, \text{ not member of } B = \text{group } A$$
$$01 = \text{not member of } A, \text{ member of } B = \text{group } B$$
$$00 = \text{not member of } A, \text{ not member of } B = \text{group } C$$

Note that a value of 1 for more than one of these variables is not possible, because it would indicate membership in more than one group. The groups must be mutually exclusive. A second feature of this coding system is that, for $g$ groups, it is necessary to use $g - 1$ dichotomous variables to completely code group membership.

## THE CASE OF TWO GROUPS

Let us suppose for the purpose of illustration that we have two groups of subjects, males and females, who have taken two tests, a measure of ability to perceive the spatial relations among objects ($SA$) and a measure of verbal fluency ($VF$). In this situation predicting which subjects are males and which are females is no problem unless we are not permitted to see the subjects themselves. However, for the latter case or for describing a composite of the two tests that shows the greatest difference between the two groups we shall need to perform a discriminant analysis.

If we form an ordinary scatter plot (variables as axes) of the pairs of test scores for the members of one group, the result is an elliptical figure. The scatter plot for the other group will also be an elliptical figure, but probably it will not be located in exactly the same place as the one for the first group. A pair of possible scatter plots for males and females on the two tests is shown in Figure 8-2. The distributions on the test axes show the abilities of the two tests, taken singly, to discriminate between the two groups. Obviously, there is substantial overlap between the groups on each variable, but females are seen to be superior on verbal fluency and males superior on spatial ability.

The problem now is to find some composite or combination of the two tests that will minimize the overlap between the two groups in terms

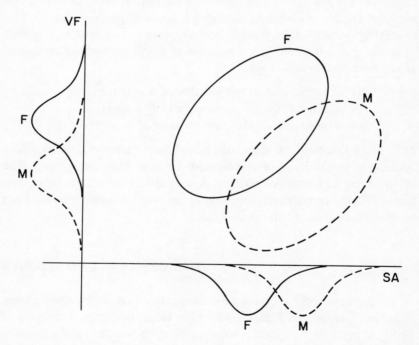

*Figure 8-2   Scatter plots for two groups on two variables. Univariate distributions are given on the axes.*

of their scores on that composite. Because the scatter plots are bivariate distributions of scores for the groups, what we are really trying to do is to define some function of the two tests that will reflect the minimum overlap between the ellipses. This has been done for the situation we are considering by drawing in the line marked *C* in Figure 8-3. A line such as this that passes through the two points where the ellipses intersect obtains the same result for the two-variable case that we got with the optimum cutting score for a single variable. By using a decision rule which says that people on one side of line *C* are predicted to be males and people on the other side are predicted to be females we will get the highest number of correct predictions. We will make some errors, but our batting average will be higher than for any other prediction strategy. Comparison of the overlap for the single variables in Figure 8-2 with that for the composite in Figure 8-3 illustrates the gain in predictive accuracy (or reduction in

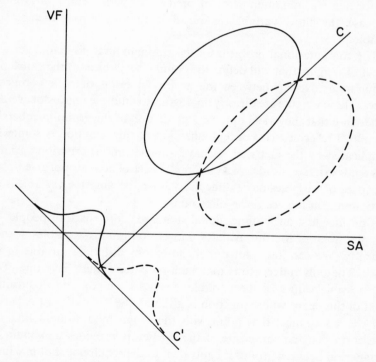

*Figure 8-3  Optimum discrimination between two groups using two variables. The distributions on composite C' have the smallest possible overlap.*

errors of classification) that would be realized by using the best combination of the two variables.

Actually, line *C* does not give us the results we need directly, because it is not a composite of the two variables that we can specify. The composite we need is a line that passes through the origin and which is perpendicular to line *C*. There is one and only one line that satisfies both of these conditions. If we project all the individual points onto this line (labeled *C'* in the figure), the distribution of scores for the two groups on the *C'* composite will have less overlap than would be the case for any other line. A discriminant analysis determines weights that will yield scores on the *C'* composite. Each person has a single discriminant score that places him at a specific location on the composite. This is the same type of result we obtained in multiple regression analysis, where we determined a composite of two variables subject to the restriction that it have maximum relationship (or minimum error of prediction) with a third variable. In this case the third variable is our dichotomous group membership variable.

The computational problem for discriminant analysis is to determine the set of weights that will define the composite, subject to the restriction of minimum overlap between the groups in terms of their composite scores. These weights are directly analogous to multiple regression weights in the sense that they yield optimum prediction of the group membership variable. The geometric conceptualization of this situation is somewhat complicated by the fact that we have a dichotomous criterion variable. This tends to make the idea of a vector in a space of people or of an ellipsoid in a space of variables more difficult to grasp, because the variable is not continuous but the vector or ellipsoid is.

Consider first an *N*-dimensional standard score space of people. We can plot vectors for our two (or more) continuous variables without difficulty. We can also plot our dichotomous criterion variable in this space. The only difference is that each individual must have one of the two possible values of the variable. Even so, we can still determine a point in this space whose location is given by the "scores" of the people, and we can connect that point with the origin by a straight line. Our restriction that the composite of the predictors provides maximum discrimination between groups is equivalent to the requirement of maximum correlation between the composite and the criterion or group membership variable, and this in turn means minimum angle between the composite vector and the group membership vector. The discriminant function weights are the coordinates of the composite in the space of the predictor variables. They are raw score regression weights in the same sense that the term was used in Chapter 6.

The analogy between discriminant function and multiple regression may be carried one step further. In Chapter 6 we saw that the $\beta$ weights could be interpreted as the slopes of the least-squares regression surface with reference to the predictor variable axes when individuals were plotted as points in a space of variables. The scatter plot of the individuals'

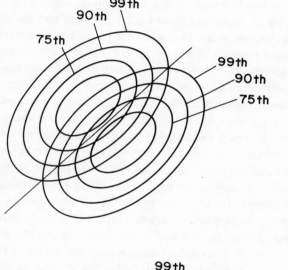

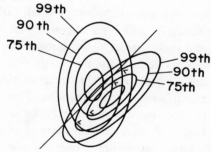

*Figure 8-4  Centours showing that unless the scatter plot for each group is the same shape the centours will not fall on a straight line.*

points in discriminant function analysis is somewhat different from the one we get in multiple regression because the criterion is dichotomous. However, the principle still holds. The scatter plot of points will contain one ellipsoid for each group, as shown in Figure 8-3, and the least-squares regression surface will be located by its weights in such a way as to provide the most accurate prediction of the criterion score for each individual.

The reader should be aware by now that discriminant function analysis does not result in dichotomous predictions of group membership. Rather, the weights, when multiplied by the standard scores for each individual, result in a score along a continuous composite. All that is necessary, as we can easily see from Figure 8-3, is to find the optimum cutting score on the composite. This score will be the one where the errors of classification are fewest. Finding this score provides the answer to the basic problem for which we might use a discriminant function analysis.

There are some necessary conditions that must be kept in mind when working with discriminant analysis. The shape and orientation of the ellipse for each group are a function of the variances of the variables and their covariances. Unless a variable has the same variance in each group and the covariances between variables are the same from group to group, the ellipses will be of different shapes and orientations. The reason these features of the ellipses must be the same is found in a fuller description of bivariate distributions. In an ordinary univariate distribution we can mark points corresponding to the various percentiles of the distribution. These percentage points, in standard score form, are the entries in the familiar $Z$ or unit normal deviate table. For example, we know that 95% of the cases in a normal distribution fall within $\pm 1.96$ units of the mean.

The situation, although somewhat more complex, is similar for bivariate and higher-order distributions. The scatter plot of people is the bivariate equivalent of the univariate frequency distribution. However, a point may deviate from the center of the bivariate plot through 360° rather than in one of two directions. Therefore, the percentile points of univariate distributions must be replaced by percentile rings in the bivariate case. This may be seen quite clearly if we assume that there is a precisely zero relationship between a pair of normally distributed standardized variables. The scatter plot would be a perfect circle. If we were to place a normal curve over the circle perpendicular to it, with pencils marking off the percentiles, and spin the normal distribution through

360°, the pencils would mark off a set of concentric circles corresponding to the percentile lines. Thus the circle with a radius of 1.96 would enclose 95% of the distribution, and other circles would correspond to other percentiles. A review of Figure 2-12 may help the reader visualize the situation.

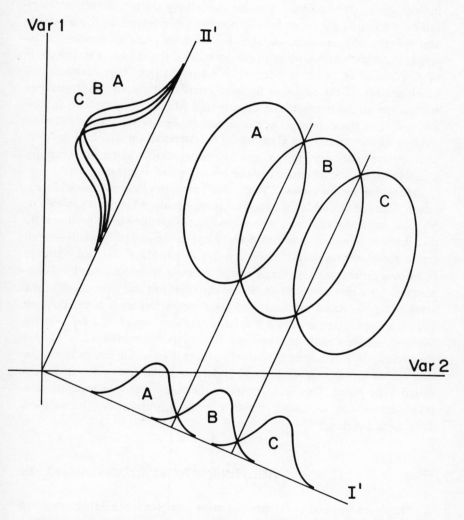

*Figure 8-5    Scatter plots of 3 groups for 2 variables, showing the first and second discriminant functions. The second composite must be independent of the first.*

The percentile contours or *centours* are slighly different when the variables are related. They then appear as a set of concentric ellipses in the bivariate distribution. Figure 8-4a contains the distributions and some of the centours for two groups such as those discussed above. Because the two distributions are equivalent (variances and covariance the same in both groups) the points where the two ninety-ninth centours intersect fall on a line that also contains the points of intersection of the ninetieth and seventy-fifth centours as well. Figure 8-4b presents a situation in which the distributions are not equivalent, and the points of intersection of the various centours do not fall on a straight line. This situation is a manifestation of the fact that the mathematics of discriminant analysis is based on an assumption that these distributions are equivalent. If they are not, then the definition of the optimum line of separation between the two distributions (line $C$ in Figure 8-3) depends on which of the centour intersections are chosen, and the separation will be differentially effective for different portions of the bivariate distributions.

**More than two variables.**   When there are more than two variables on which data are available for two groups the situation becomes somewhat, but not markedly different. For a case of $n$ variables each multivariate distribution becomes an $n$-dimensional hyperellipsoid (a football-shaped figure for three variables). Each group's distribution still has centours (now in $n$ dimensions), some of which intersect with the same level centours of the other distribution. If the distributions are equivalent in the sense of having equal variances and covariances, the areas of intersection fall on a hyperplane of $n - 1$ dimensions. (This can readily be seen for the three-dimensional case because the intersections fall on a plane.) Regardless of the number of variables, there is only one line in the space that passes through the origin and is perpendicular to the $n - 1$ dimensional hyperplane. This line represents the composite that gives the discriminant scores, scores for which the two groups show the smallest amount of overlap.

## MULTIPLE DISCRIMINANT ANALYSIS

The situation becomes somewhat more complex when data from more than two groups are being analyzed. Because it is necessary to use $g - 1$ dichotomous variables to code group membership for $g$ groups, the problem becomes one of canonical analysis rather than multiple regression analysis. That is, with two or more variables and three or more groups we have a set of predictor variables and a set of dichotomous group membership criterion variables. When this is the case, there are either $n$ (the number of variables) or $g - 1$ (whichever is smaller) different and independent discriminant composites, each corresponding to one of the possible canonical correlations.

Let us say that we have collected data on two variables for three groups of subjects. A multiple discriminant analysis would determine two composites. The first composite would be defined in such a way as to make maximum discriminations among the groups using all of the information. Unless two of the groups are identical (which would cause linear dependence), this composite will not be able to exhaust the sources of differences among the groups. The second composite then uses the residual information, that which is independent of the first composite, to make further distinctions among the groups. Scores on the second composite maximize the differences among the groups after information from the first composite is removed.

Figure 8-5 illustrates for our two-variable three-group example how the second (and succeeding) discriminant functions for larger problems would appear geometrically. The first dimension of discrimination is found such that there is minimum overlap in the distributions of first discriminant scores. In our example the position of the second dimension is fixed, because we can have only two dimensions that must be independent. The figure shows that each group has a distribution of scores on this second function also. If we had more variables and groups, the second discriminant function would be located to minimize the overlap of these distributions, subject to the restriction of independence, and we could continue to define additional dimensions up to the smaller of $n$ and $g - 1$.

## INTERPRETING DISCRIMINANT ANALYSES

We have seen that discriminant function analysis is really a special case of multiple regression or canonical analysis. Because this is true, the composites are open to the same general type of interpretation. That is, the discriminant weights are the coordinates of the composites and give their locations with respect to the variables. Likewise, the component loadings are the correlations of the variables with the composites and may be used to interpret the composites in terms of their relationships with the observed variables.

The fact that discriminant analysis involves discrimination among groups of individuals permits us to use another approach to interpretation. It is possible to calculate a score for each individual on each composite and to compute the mean of each group on each composite. If the groups have been selected in such a way that there are meaningful a priori differences among them, then their relative positions on the discriminant composites may aid in interpreting the composites. We might, for example, have obtained measures on five ability tests from people employed in ten different occupations. For such a hypothetical situations we might

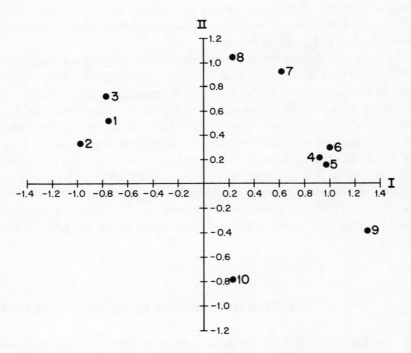

*Figure 8-6   Plot of group means on the first two discriminant functions for ten hypothetical occupational groups from Table 8-1.*

have found two of the five possible discriminant composites to be statistically significant. Determining composite scores and computing group means might yield results such as those shown in Table 8-1. Because the discriminant composites are by definition independent, they may be viewed as a pair of orthogonal axes, and we can represent our results graphically by using the means of the groups as coordinates of the groups on these axes. This has been done in Figure 8-6, where it may be seen that the occupations fall into meaningful groups. (Because these are hypothetical data, the reader may rest assured that the "meaning" is at the whim of the author. Real data are seldom this clear in their interpretation.)

Let us assume that the variables involved in our hypothetical analysis are measures of physical strength and measures of intelligence. Looking at Table 8-1 and Figure 8-6 we see that the first composite seems to distribute the occupations from a high group (elephant trainer, carpenter, bricklayer, auto mechanic) that requires physical strength down to a low group (secretary, file clerk, librarian) that does not require physical strength. Interpretation of this as a "physical strength" dimension might be confirmed by finding that the tests with high positive loadings on it involve strength.

**TABLE 8-1**

*Group Means on 2 Discriminant Functions for*
*10 Hypothetical Occupations.*

| Occupation | Mean on Composite I | Mean on Composite II |
|---|---|---|
| 1. Secretary | −.75 | +.52 |
| 2. File clerk | −.98 | +.33 |
| 3. Librarian | −.77 | +.72 |
| 4. Auto mechanic | +.92 | +.22 |
| 5. Bricklayer | +.97 | +.16 |
| 6. Carpenter | +1.00 | +.30 |
| 7. Surveyor | +.62 | +.93 |
| 8. Accountant | +.23 | +1.05 |
| 9. Elephant trainer | +1.30 | −.39 |
| 10. Berry picker | +.23 | −.78 |

On the second discriminant composite, the high-scoring groups (accountant, surveyor) are those occupations that seem to require more intelligence, in contrast to the job of berry picker at the low end of the scale. Again, finding high positive loadings for the appropriate predictor variables (the intelligence measures) would help confirm this interpretation.

It is a unique (and perhaps comforting) feature of discriminant analysis that there are two courses one can take in arriving at an interpretation of the data. The other multivariate procedures described in this book provide only one route for developing an explanation of the meaning of an analysis, the relationships between the variables and the composites. In discriminant analysis we have meaningful information on group membership that can add to the information from the variable-composite relations. In a sense, each source of information permits a cross-check on the interpretation derived from the other. One should be wary of an interpretation from one source of information that is not confirmed by the interpretation from the other.

## MATHEMATICAL CONSIDERATIONS

Although discriminant analysis can be viewed as a special type of multiple regression or canonical analysis as described above, the correlation matrix is not ordinarily used in its computations. Instead, we must define a new type of matrix which we will call a *deviation crossproducts matrix*. A deviation crossproducts matrix is formed by taking the sums of squares and sums of crossproducts of deviation scores for each of the variables across a given group of individuals. In this case we define our deviation scores as deviations from means found from the total group data rather than from the subgroup means. That is, $x_i = X_i - \overline{X}$ rather than $x_{jc} = X_{jc} - \overline{X}_j$. If we have three variables ($X_1$, $X_2$, $X_3$) and three groups (of size $n_1$, $n_2$, and $n_3$) of subjects, we can form one matrix, a total crossproduct matrix $\mathbf{G}$, in the following way (note that this is equivalent to $\mathbf{G} = \mathbf{D'D}$, where $\mathbf{D}$ is a deviation score matrix as discussed in Chapter 4).

$$N = n_1 + n_2 + n_3$$

$$\mathbf{G} = \begin{bmatrix} \sum_{i=1}^{N} x_1{}^2 & \sum_{i=1}^{N} x_1 x_2 & \sum_{i=1}^{N} x_1 x_3 \\ \sum_{i=1}^{N} x_2 x_1 & \sum_{i=1}^{N} x_2{}^2 & \sum_{i=1}^{N} x_2 x_3 \\ \sum_{i=1}^{N} x_3 x_1 & \sum_{i=1}^{N} x_3 x_2 & \sum_{i=1}^{N} x_3{}^2 \end{bmatrix} \tag{8.1}$$

This matrix contains the total sums of squares of the variables in its main diagonal and the total sums of deviation crossproducts among the variables as its off-diagonal elements.

Actually, **G** is not the matrix we really need. From analysis of variance we know that it is possible to partition total variance into two parts, variance among groups and variance within groups. We can do the same thing with a crossproducts matrix to obtain a matrix **A**, which is a matrix of crossproducts and sums of squares among groups, and a matrix **W**, a matrix of pooled crossproducts and sums of squares within groups. These matrices are found in exactly the same way we found the sums of squares between and within groups when computing the correlation ratio in Chapter 2, except that now we have three variables to contend with, and we must account for the relationships among variables. Thus the **W**-matrix is formed in the following general way:

$$\mathbf{W} = \begin{bmatrix} \sum_{j=1}^{G} \sum_{i=1}^{N_g} (X_{1ij} - \bar{X}_{1j}) & & \\ \sum_{j=1}^{G} \sum_{i=1}^{N_g} (X_{1ij} - \bar{X}_{1j})(X_{2ij} - \bar{X}_{2j}) & \sum_{j=1}^{G} \sum_{i=1}^{N_g} (X_{2ij} - \bar{X}_{2j})^2 & \\ \sum_{j=1}^{G} \sum_{i=1}^{N_g} (X_{1ij} - \bar{X}_{1j})(X_{3ij} - \bar{X}_{3j}) & \sum_{j=1}^{G} \sum_{i=1}^{N_g} (X_{2ij} - \bar{X}_{2j})(\bar{X}_{3ij} - \bar{X}_{3j}) & \sum_{j=1}^{G} \sum_{i=1}^{N_g} (X_{3ij} - \bar{X}_{3j})^2 \end{bmatrix}$$

The **A**-matrix is found in an analogous way, or, remembering the additive property of sums of squares (i.e., $SS_T = SS_B + SS_W$) we can say that $\mathbf{G} = \mathbf{A} + \mathbf{W}$ and that $\mathbf{A} = \mathbf{G} - \mathbf{W}$.

The problem is to find a vector (**c**) of weights for the variables that will maximize the ratio

$$\lambda_k = \frac{\mathbf{c}_k' \mathbf{A} \mathbf{c}_k}{\mathbf{c}_k' \mathbf{W} \mathbf{c}_k} \tag{8.2}$$

for the $k$th discriminant function. This amounts to maximizing our ability to discriminate among the groups on the basis of a composite of the variables defined by the vector of weights. The solution of the matrix equation

$$(\mathbf{W}^{-1}\mathbf{A} - \lambda_k \mathbf{I})\mathbf{c}_k = 0 \tag{8.3}$$

provides the necessary vector of weights in terms of deviation scores. However, it is necessary to standardize these weights so that the resulting discriminant scores are comparable from one composite to another, thus making them essentially equivalent to $\beta$ weights. This is accomplished by finding the total variance ($V_k$) of discriminant scores (across all subjects in all groups) for a given composite by the formula

$$V_k = \mathbf{c}_k' \left( \frac{1}{N-1} \right) \mathbf{G} \mathbf{c}_k \tag{8.4}$$

and by defining **P** as a diagonal matrix of the square roots of $1/(N-1)$ times the diagonal elements of **G**. Thus the elements of **P** are the standard deviations of the variables for the total group. The desired vector of standardized weights ($\mathbf{b}_k$) is given by

$$\mathbf{b}_k = \mathbf{P} \mathbf{c}_k \sqrt{\frac{1}{V_k}} \tag{8.5}$$

The discriminant score for any individual on a particular composite is found by multiplying his standard scores on the variables by the vector of weights for each of the possible composites. These can be collected into a matrix of discriminant weights, **B**, which has the weights for the $n$ or $g - 1$ discriminant functions as its columns. (Note that standard scores are computed in terms of the total group means and standard deviations.)

Up to this point we have discussed discriminant analysis as a technique for predicting group membership. It is also possible to use the producedure for external factor analyses similar to those described for multiple regression and canonical analysis. All that is necessary to generate the discriminant components is to multiply the correlation matrix for the total sample by the discriminant weights. For a particular discriminant function, the *i*th, the structure, which is the vector of correlations between the variable scores and the scores on the *i*th discriminant function, is given by

$$s_i = Rb_i$$

The matrix equation

$$S = RB \tag{8.6}$$

gives the total structure for all the discriminant functions.

The discriminant components, which are given in the columns of the structure matrix, may be used as outlined above for describing dimensions that differentiate the groups. The first discriminant component is the dimension on which the groups show maximum differences and can be interpreted as the "factor" on which the groups are most different. Succeeding components have similar interpretations, so that the whole solution can give a picture of the structure of these differences.

The group means described earlier as an alternate method for interpreting discriminant functions can be computed either by the brute-force method of determining discriminant scores for all individuals and taking group averages or by the simpler process of multiplying the deviation of the group means from the total sample means on each variable by the matrix of partially standardized weights. If we arrange the means of the total group in the form of a vector, $\bar{X}$, and do the same for the means of the *j*th group, $\bar{X}_j$, then the deviations are given by

$$\bar{X}_j^* = \bar{X}_j - \bar{X}$$

and the means for the group in the discriminant space are

$$\bar{X}_{d_j} = \left( C \sqrt{\frac{1}{V}} \right)' \bar{X}_j^* \tag{8.7}$$

These means can be plotted in the discriminant space that has the discriminant functions as its axes to show the relationships between the groups graphically, as illustrated in the previous section.

The statistical significance of a discriminant function is estimated in much the same way as for a canonical analysis. First, we arrange the lambdas ($\lambda_i$) in order from smallest to largest and define Wilks' lambda for the complete set of discriminant functions as

$$\Lambda = \prod_{i=1}^{p} \left( \frac{1}{1 + \lambda_i} \right) \tag{8.8}$$

where $p$ is the number of discriminant functions (the smaller of $n$ or $g - 1$). When this value is placed in the equation

$$\chi^2 = - \left( N - \frac{n + g}{1} \right) \log_e \Lambda, \tag{8.9}$$

the $\chi^2$ with $n(g - 1)$ degrees of freedom is a test of whether the discrimination among the groups is significantly better than chance. The significance of the $k$th discriminant function may be tested by defining

$$\Lambda_k = \prod_{i=1}^{p-k} \left( \frac{1}{1 + \lambda_i} \right).$$

Inserting $\Lambda_k$ into equation (8.9) and using $df = (n - k)(g - k - 1)$, we have the needed test. Generally, only the first few discriminant functions are of interest or statistically significant, which simplifies the problem of plotting group means for interpretation.

## AN EXAMPLE

For our example of the use of discriminant function analysis we return once again to the occupational data presented in the two previous chapters. The 130 individuals who provided the data were of two types. Sixty-five of them were judged successful employees, and 65 were judged unsuccessful. Discriminant function analysis can be applied to these data to determine the weighted linear composite of the predictor variables that most accurately separates the individuals into successes and failures.

The two groups of 65 were further divided randomly into groups of 50 subjects for development of the discriminant function and 15 for cross-validation of the results. Thus the linear composite was developed on a group of 100 (50 successes and 50 failures) and cross-validated on a group of 30.

The development analysis yielded the results shown in the left-hand portion of Table 8-2. There is, of course, only one possible discriminant function, because there are only two groups. The weights that give the optimum discrimination show that the composite is composed primarily of variance from variable 10, with some contributions from variables 3, 4, 7, 13, 16, and 18. People who were judged successful tended to be those who owned their homes and, to a lesser extent, had less education, higher GPA's, held fewer offices, had more older siblings, belonged to fewer organizations, and had lower aspirations. The composite thus described

**TABLE 8-2**

*Discriminant Function Analysis and Cross-Validation Using 18 Variables from Table 6-2 and Accept-Reject Grouping.*

| | Development | | Cross-validation Loadings |
|---|---|---|---|
| | Weights | Loadings | |
| 1 | −.002 | −.253 | −.007 |
| 2 | .082 | .241 | −.073 |
| 3 | −.148 | −.353 | −.563 |
| 4 | .126 | −.048 | .118 |
| 5 | .000 | −.059 | −.102 |
| 6 | −.019 | −.411 | −.030 |
| 7 | −.121 | −.225 | −.320 |
| 8 | −.089 | −.267 | −.611 |
| 9 | −.038 | −.530 | −.138 |
| 10 | .343 | −.040 | .717 |
| 11 | −.006 | −.092 | .086 |
| 12 | −.054 | .025 | .276 |
| 13 | .165 | −.033 | .312 |
| 14 | −.007 | −.247 | .190 |
| 15 | −.022 | −.247 | .074 |
| 16 | −.103 | −.236 | −.235 |
| 17 | −.008 | −.335 | −.154 |
| 18 | −.161 | −.365 | −.294 |

$R = .51$      Centroids      $R = .02$

$\lambda_i = .358$      Group 1   −.511

$\Lambda = .7356$      Group 2   +.511

$\chi^2 = 27.23$

$p < .08$

had a multiple correlation with the dichotomous variable of group membership of .51. The eigenvalue for this function ($\lambda_i$) was .358, yielding a $\chi^2$ of 27.23 with 18 degrees of freedom, which is not quite statistically significant. At first glance the results of the developmental analysis, although not overwhelming, might seem to justify use of these variables for the prediction of group membership. However, a more accurate index of their value is revealed in the second part of the table, where we see that when the weights were applied to a new group, the correlation between the composite thus defined and the group membership variable is .02. This level of relationship is hardly sufficient to justify using the discriminant weights to predict job success, and the shrinkage of the correlation from .51 to .02 is a good indicator of the possible effects of sample specific covariation.

An additional point should be made about cross-validation in discrimination analysis. As is the case in canonical analysis, the cross-validation loadings often show substantial similarity to the development loadings, even when the cross-validation is poor. This is to be expected, because the discriminant weights are constant across the two analyses. The marked dissimiliarities in the present example (e.g., variable 10) are an example of how badly the cross-validation worked for these data. The much more normal situation is that found in Chapter 7, where the same 30 individuals served as the cross-validation group (although those results are still not as good as prior research might lead one to expect).

Note that although raw scores may be used in the development analysis, the cross-validation requires that standard scores be computed. However, the means and standard deviations for the total cross-validation group are used to compute these standard scores. This preserves the subgroup mean differences, which are the basis of discriminant analysis. Of course, if the deviation score weights, $\mathbf{c}$, from equation (8.3) are available, they may be used to compute composite scores. However, the results in terms of discriminant loadings will not be comparable with the development analysis.

## SUMMARY

Discriminant function analysis is a special case of canonical analysis in which the variables in one set are dichotomous dummy variables generated to identify group membership. When there are $g$ groups, $g - 1$ dummy, variables are required. The discriminant analysis defines composites of the predictor variables that have maximum canonical correlations with composites of the dummy group membership variables. The effect of this type of analysis is to maximize the accuracy of prediction of group membership. This is accomplished by finding composites of the predictor set such that the composite score distributions for the groups show minimum overlap. As is the case with the other procedures discussed in Part II, the composites are defined by the standard discriminant weights and can be described in terms of their discriminant loadings, which are the correlations of the predictor variables with the composites. Also, as is shown by the example at the end of the chapter, it is necessary to cross-validate the discriminant weights to assess their value for predicting group membership in new samples.

# PART III
## INTERNAL FACTOR ANALYSIS

# 9

# CLUSTER ANALYSIS

In Part III we turn our attention to the internal structure of a set of variables. The procedures presented in this and the following chapters focus on describing the structure of a set of variables without reference to any variable that is not in the set. We begin with a procedure called *cluster analysis*, which has sometimes been referred to as "a poor man's factor analysis" because the computations involved in the early versions of this technique are simpler than those of early forms of factor analysis. Our discussion of cluster analysis centers on the simplest form of the procedure, because it has important differences with factor analysis, differences that become less clear in more advanced clustering methods. Also these later developments are given a thorough treatment in a recent book by the men most responsible for their development (Tryon and Bailey, 1970).

After presenting cluster analysis as one logical alternative for analyzing the internal structure of a set of variables, we turn to traditional factor analysis. In Chapter 10 we present the general model and discuss the logic of this class of procedures. Chapter 11 is devoted to a discussion of such controversial topics in factor analysis as the proper number of factors

**227**

and the choice of a particular solution. Chapter 12 includes some suggestions on research methodology and the interpretation of factors, a discussion of some special problems, and an overview of factor analysis.

## CLUSTER ANALYSIS OF VARIABLES

We have seen that a positive correlation between two variables means
that their vectors make an angle of less than 90° with each other in a scatter
plot of the variables. When several variables are positively correlated
their vectors form a group or cluster. Variables that are highly correlated
form a tighter group than do variables with lower correlations. Cluster
analysis of a set of variables provides a way of identifying those variables
that tend to cluster in this way by applying a simple set of decision rules
to a correlation matrix. Although there are several ways in which cluster
analysis may be performed by applying different sets of rules, the procedure we present is Tryon's modification of the Holzinger–Harman
method, as described by Fruchter (1954). It is mathematically simple and
provides a means of separating the variables into nonoverlapping (but
correlated) clusters.

The central feature of cluster analysis is the $b$ coefficient or "coefficient
of belonging." This index, which is calculated each time a new variable is
added to a cluster, is used to indicate the strength of association among
the variables in the cluster. The $b$ coefficient is defined as the ratio of the
mean of the correlations among variables in the cluster to the mean of
the correlations of the variables in the cluster with those variables not
in the cluster, or

$$b = \frac{\bar{r}_{ii}}{\bar{r}_{io}} \qquad (9.1)$$

If we have 10 variables and 4 of them are included in the cluster, $\bar{r}_{ii}$ is
the mean of the 6 correlations among the 4 included variables and $\bar{r}_{io}$
is the mean of the 24 correlations between the 4 included variables and
the 6 not-included variables.

**Performing a cluster analysis.**   A cluster analysis is performed by following this set of decision rules:

1. Find the largest off-diagonal element in the correlation matrix (the highest correlation between two variables). These two variables form the nucleus of the cluster.
2. Add each of the remaining variables to the cluster in turn and calculate the $b$ coefficient for the cluster with that variable included.
3. Add to the cluster the variable whose inclusion yields the highest $b$ coefficient for the new cluster (of three variables).
4. Repeat steps 2 and 3 for a fourth variable, adding to the cluster that variable which yields the highest $b$ coefficient.
5. Continue adding variables by the above procedure until there is a sharp drop in the $b$ coefficient or until the $b$ coefficient falls below some predetermined value. What constitutes a sharp drop or a minimum acceptable value of $b$ is a matter for the individual investigator to decide. If a loose clustering is satisfactory, a low criterion value for $b$ may be used. The higher the minimum $b$, the tighter the cluster will be.
6. When the decision is reached that the first cluster is complete, a new cluster may be started by searching among the variables that have not been clustered for the most highly correlated pair and proceeding as above, being careful not to include already clustered variables in the new cluster.
7. Variables are added to the second cluster until the $b$ coefficient for that cluster becomes too low.
8. Additional clusters may be formed from among the remaining variables until all variables have been placed in one or another cluster or until there is no pair of variables remaining that yields a satisfactory $b$ coefficient, at which point clustering is complete. Note that it is quite possible for some variables to remain unclustered, particularly when tight clusters are desired and a high minimum $b$ is set.

One of the important differences between cluster analysis and factor analysis is that a variable is assigned to only one cluster, whereas factor analysis, as we shall see in the next chapter, breaks up the variance of a variable into several additive parts. This means that cluster analysis is the proper technique to use when the analysis goal is to build up something, for example, a scale, from several smaller somethings, for example, items. On the other hand, factor analysis is properly used when the objective is to break down the variance of a scale or item into independent parts. We may say that cluster analysis is a method of *synthesis*, whereas factor analysis is a method of *analysis*. Both have their uses, but it is a common mistake to consider clustering procedures to be poor relatives of the more elegant factoring ones.

An example of a research problem in which cluster analysis is the appropriate technique is in the development of homogeneous scales from a set of items. A clustering procedure such as the one outlined above will result in nonoverlapping groups of items. The correlations among the items within the group will be high, and correlations with other items will be low. Factor analysis would be inappropriate for this phase of research, because it would assign a portion of the variance of each item to each of several factors. This confuses the issue of assigning items to particular scales. In studies in which factor analysis has been used in this way each item is usually assigned to the factor scale with which it is most closely associated, a result which is not appreciably different from a cluster analysis.

Cluster analysis has another important difference with factor analysis. Although the procedures of factor analysis guarantee that the factors initially will be uncorrelated with each other, there almost certainly will be nonzero correlations among clusters. This feature is not of particular importance except that (1) it makes possible further analyses of the clusters (such as a factor analysis!) and (2) there is some disagreement about the basic nature of the universe. Some investigators, notably Thurstone (1947), Tryon (1958), and Cattell and Dickman (1962) have argued vigorously that the world is made up of correlated dimensions (but preferably derived by factor analysis). Others, for example, Reyburn and Raath (1949), have argued that one should use uncorrelated dimensions to understand the world. However, regardless of one's philosophical position on this matter, we can say with some assurance that scales developed using cluster analysis will be easier to score and make more sense to the general user than will factor scales, because the latter require breaking up the variance of single items.

Some researchers who use factor analysis have attempted to circumvent the problem of breaking up item variance by assigning each item to the factor with which it shows highest association. This procedure is acceptable if one realizes that he no longer has a factor analysis when he is finished. Some of the mathematical properties of factor analysis, such as independence of the factors and an exact decomposition of the correlation matrix, will be destroyed. We may then ask why a cluster analysis was not done in the first place.

It is quite easy to develop scale scores for subjects when the scales have been developed by cluster analysis. The score for a particular cluster scale is simply the sum of the scores on the several items in the cluster. Given these scale scores, it is a simple matter to calculate the correlations among the clusters. A factor analysis can then be performed on the scale correlation matrix to obtain a description of the set of scales in terms of independent dimensions. This aids in identifying the content of each of the cluster scales.

A problem can arise in performing a cluster analysis. Negative relationships among some of the variables can cause difficulties in finding satisfactory clusters because they may deflate the value of $\bar{r}_{io}$. This will tend to force too many variables with low correlations into the cluster. Suppose, for example, that we have 10 variables and that the correlation matrix shows that variables 1 and 2 are highly correlated, variables 3, 4, and 5 have low positive correlations with 1 and 2 and with each other, and variables 6 to 10 have high negative correlations with 1 and 2 and moderate negative correlations with 3 to 5. The result would probably be that variables 1 to 5, which have a low average correlation among them, would be clustered together. We would probably consider this result to be unsatisfactory.

One way to overcome this problem is to consider each variable as being *bipolar*. This means that we would assume that there is a rational opposite to the trait a given variable is measuring. Had we constructed the variable to measure this rationally opposite trait, its correlations with the other variables would have been positive. Thus we may make all correlations positive for the purpose of doing the cluster analysis and avoid the detrimental effects of high negative correlations on our clusters. When this is done we often find that variables that are negatively correlated appear in the same cluster. Clusters of this type are called bipolar clusters, meaning that there are variables at both ends of the continuum the cluster represents. Another way to look at this result is to say that there are two clusters that are highly negatively correlated.

The presence of bipolar clusters requires caution on the part of the investigator, both in the interpretation of the clusters and in the computation of cluster scores. The computer program being used may have been written to reverse negative correlations before clustering the variables. The investigator should check the clusters against the original correlation matrix to ensure that bipolar clusters are identified as such before proceeding further. When cluster scores are to be computed from bipolar clusters, the scales for variables at one pole of the cluster must be reversed before cluster scores are determined.

## AN EXAMPLE

Data from a study by Fleishman and Hempel (1954) are used to illustrate the procedures involved in a cluster analysis. The variables of this study were 15 manual dexterity tests that were administered to 400 subjects. The matrix of correlations among the variables is given in Table 9-1 (only the lower triangle of this symmetric matrix is given). These same data

**TABLE 9-1**

*Correlations among 15 Manual Dexterity Tests (from Fleishman and Hempel, 1954).*

|    | 1    | 2    | 3    | 4    | 5    | 6    | 7    | 8    | 9    | 10   | 11   | 12   | 13   | 14   | 15   |
|----|------|------|------|------|------|------|------|------|------|------|------|------|------|------|------|
| 1  | 1.00 |      |      |      |      |      |      |      |      |      |      |      |      |      |      |
| 2  | .45  | 1.00 |      |      |      |      |      |      |      |      |      |      |      |      |      |
| 3  | .41  | .38  | 1.00 |      |      |      |      |      |      |      |      |      |      |      |      |
| 4  | .47  | .39  | .48  | 1.00 |      |      |      |      |      |      |      |      |      |      |      |
| 5  | .46  | .37  | .42  | .49  | 1.00 |      |      |      |      |      |      |      |      |      |      |
| 6  | .45  | .40  | .38  | .45  | .44  | 1.00 |      |      |      |      |      |      |      |      |      |
| 7  | .43  | .44  | .30  | .41  | .39  | .50  | 1.00 |      |      |      |      |      |      |      |      |
| 8  | .32  | .18  | .27  | .25  | .33  | .41  | .32  | 1.00 |      |      |      |      |      |      |      |
| 9  | .36  | .40  | .24  | .38  | .28  | .40  | .46  | .28  | 1.00 |      |      |      |      |      |      |
| 10 | .34  | .27  | .28  | .29  | .31  | .35  | .30  | .30  | .40  | 1.00 |      |      |      |      |      |
| 11 | .29  | .23  | .27  | .25  | .29  | .36  | .40  | .33  | .39  | .35  | 1.00 |      |      |      |      |
| 12 | .32  | .33  | .26  | .31  | .30  | .42  | .40  | .34  | .43  | .37  | .50  | 1.00 |      |      |      |
| 13 | .21  | .26  | .15  | .20  | .27  | .29  | .28  | .15  | .28  | .21  | .32  | .42  | 1.00 |      |      |
| 14 | .21  | .21  | .20  | .19  | .20  | .22  | .25  | .16  | .33  | .13  | .30  | .32  | .30  | 1.00 |      |
| 15 | .29  | .27  | .26  | .28  | .35  | .39  | .35  | .25  | .37  | .26  | .39  | .49  | .40  | .32  | 1.00 |

are used in later chapters to illustrate various factoring procedures and to allow comparison of clustering and factoring results.

As we noted earlier, the value chosen for $b$ controls the tightness of the clusters to be obtained. The author prefers to use values of 2.0 or greater for $b$, although some authors (e.g., Fruchter, 1954) recommend lower values. The present set of data illustrates quite well the fact that it may be necessary to perform several analyses using different values for $b$ before a satisfactory solution can be obtained.

First, a cluster analysis was performed using 2.0 as the critical value for $b$. No clusters were found. In fact, no pair of variables yielded a $b$ coefficient of 2.0. A preliminary hand computation showed that a much lower $b$ was necessary, so a critical value of 1.2 was input to the program. The results of this analysis are shown in Table 9-2. This critical value is also unsatisfactory, because all the variables were placed in a single cluster. The table shows that variables 6 and 7, which have a correlation of .50, are the first to be placed in the cluster and yield a $b$ coefficient of 1.34. The addition of the next variable, number 1, causes this value to drop slightly, but then it rises again and stays above 1.2 until all variables have

**TABLE 9-2**

*Initial Clustering of 15 Variables Using a Minimum b of 1.2.*

| Cluster | Beta Coefficient | Variable | | | | | | | | | | | | | |
|---------|------------------|---|---|---|---|---|---|---|---|----|----|----|----|---|----|
| 1 | 1.342 | 6 | 7 | | | | | | | | | | | | |
| 1 | 1.280 | 6 | 7 | 1 | | | | | | | | | | | |
| 1 | 1.314 | 6 | 7 | 1 | 4 | | | | | | | | | | |
| 1 | 1.364 | 6 | 7 | 1 | 4 | 5 | | | | | | | | | |
| 1 | 1.390 | 6 | 7 | 1 | 4 | 5 | 2 | | | | | | | | |
| 1 | 1.440 | 6 | 7 | 1 | 4 | 5 | 2 | 3 | | | | | | | |
| 1 | 1.389 | 6 | 7 | 1 | 4 | 5 | 2 | 3 | 9 | | | | | | |
| 1 | 1.321 | 6 | 7 | 1 | 4 | 5 | 2 | 3 | 9 | 12 | | | | | |
| 1 | 1.304 | 6 | 7 | 1 | 4 | 5 | 2 | 3 | 9 | 12 | 15 | | | | |
| 1 | 1.306 | 6 | 7 | 1 | 4 | 5 | 2 | 3 | 9 | 12 | 15 | 11 | | | |
| 1 | 1.350 | 6 | 7 | 1 | 4 | 5 | 2 | 3 | 9 | 12 | 15 | 11 | 10 | | |
| 1 | 1.426 | 6 | 7 | 1 | 4 | 5 | 2 | 3 | 9 | 12 | 15 | 11 | 10 | 8 | |
| 1 | 1.435 | 6 | 7 | 1 | 4 | 5 | 2 | 3 | 9 | 12 | 15 | 11 | 10 | 8 | 13 |

Minimum $b$ = 1.2.

been put in the cluster. A result such as this tells us little about the internal structure of the set.

The reader should note that the last variable (number 14) is not included in the table. Computer programs for performing this kind of cluster analysis, as a rule, do not assign the last variable to a cluster because it is then impossible to compute the $b$ coefficient. With no variables remaining outside the cluster, the denominator of the $b$ ratio is zero, and this situation causes the computer to suffer a severe case of indigestion (perhaps followed by a string of expletives, depending on the sense of humor of the systems programmer).

The results from Table 9-2 can be used to give us a clue about what will be the best value for the $b$ coefficient. Any value below 1.28 will yield only a single cluster, and any value above 1.34 will yield no clusters at all. If we split the difference and use 1.31 as our critical value for $b$ (we could just as well use 1.30), we obtain the results shown in Table 9-3. These results seem reasonably satisfactory, in that the three clusters which result provide some differentiation within the set of variables. Also, the contents of the clusters make fairly good sense. The first cluster, which is composed of the turning and placing tasks on the Minnesota Rate of Manipulation Test, seems to involve large muscle movements and eye-hand coordination. The second cluster involves eye-hand coordination in marking tasks, and the third cluster is composed of variables that test fine motor coordination.

**TABLE 9-3**
*Three Clusters Found with 1.31 Minimum b.*

| Cluster | Beta Coefficient | Variables | | | | |
|---------|------------------|-----|-----|-----|-----|-----|
| 1 | 1.342 | 6 | 7 | | | |
| 2 | 1.464 | 11 | 12 | | | |
| 2 | 1.405 | 11 | 12 | 15 | | |
| 3 | 1.454 | 4 | 5 | | | |
| 3 | 1.431 | 4 | 5 | 1 | | |
| 3 | 1.473 | 4 | 5 | 1 | 3 | |
| 3 | 1.441 | 4 | 5 | 1 | 3 | 2 |

Minimum $b$ = 1.31.

Again, we see that not all variables have been assigned to clusters. In this case, 5 of the 15 have not found a home. This is a common result and can be handled in any of three ways. One solution is to consider that the unclustered variables are extraneous to the desired result and discard them. This might be appropriate if the investigator had included a large number of diverse variables (perhaps items) with the objective of identifying homogeneous scales or groups. In instrument development research the decision might be made to keep the items in the instrument but leave them unscored for the purpose of homogeneous scales. (They might function well in empirically derived scales.) Whether to throw them out or ignore them depends upon the situation.

A second alternative is to treat the unclustered variables as members of inadequately measured clusters. If other variables like these had been included in the original set, clusters would have been formed. Thus new variables would be added to the set in subsequent research. This alternative is particularly appropriate when one is interested in a complete measurement of a domain of behavior and the present study is for exploratory and domain definition purposes.

The third way to handle unclustered variables is to assign them to the clusters with which they have the highest average correlations. Doing this means that the $b$ coefficients may drop below the desired level. It also tacitly assumes that the investigator will find the new clusters meaningful in terms of the content of the added variables. That is, there should be logical as well as statistical justification for placing the variables. However, given the rather arbitrary nature of the selection of a $b$ coefficient, a certain lack of precision is tolerable.

There are two other aspects of this particular cluster analysis that are worthy of note. First, all the correlations in Table 9-1 are positive, and none is below .13. It is this feature of the correlation matrix that makes it relatively difficult to obtain a satisfactory solution. The variables really do form one large, loose cluster, and the solution we have chosen in Table 9-3 identifies three tighter clusters within the overall one.

The second feature of these data that makes them unusual is that successive clusters contain more variables. The usual finding in a cluster analysis is that the first cluster is fairly large and the succeeding clusters contain fewer and fewer variables. Quite often a cluster analysis of a large set of variables will yield a few large clusters and several small clusters containing two or three variables. It is up to the individual investigator to decide for himself, given his data, how large a cluster must be to be worth keeping for interpretation. A similar problem arises in factor analysis.

## ALTERNATIVE STRATEGIES IN CLUSTER ANALYSIS

Up to this point our discussion has always used variables as meaning tests or other dimensions on which data have been collected. This is not the only way in which we can view the situation, and some of the alternatives may prove extremely useful in a variety of research situations.

**The data cube.** In most situations in behavioral research our data consist of responses of individual subjects or groups of subjects. A particular response may be classified according to its source (each of $i$ subjects or groups of subjects would be a source of data), the stimulus to which it was made (e.g., test item, or light intensity), and the number of times (occasions) the stimulus has been presented before. Figure 9-1 displays the organization of data set up in this way. The vertical dimention, $i$, represents subjects 1 to $N$, the horizontal dimension, $j$, represents stimuli 1 to $n$, and the depth dimension, $k$, represents occasions 1 to p. Thus a single cell of the cube, $r_{ijk}$, refers to the response of individual $i$ to stimulus $j$ on occasion $k$.

Most studies involve only a portion of the data cube. A typical cluster analysis such as the one in our example deals with only one slab of the cube, a 400-person by 15-test slab at a single point in time (usually occasion 1). The correlations computed for data of this type are generally the Pearson product-moment correlations with which we are familiar. They are found by collapsing across the subject dimension to obtain the $15 \times 15$ matrix found in Table 9-1. If we consider these to be standard scores, then "collapsing across subjects" to form the correlation matrix is equivalent to the matrix equation $\mathbf{R} = (1/N)\mathbf{Z}'\mathbf{Z}$, where $\mathbf{Z}$ is the 400 (rows) $\times$ 15 (columns) standard score matrix. However, if we can collapse across the subject dimension (i.e., compute correlations among variables), we should also be able to collapse across the other dimensions. By taking the same $400 \times 15$ data matrix and collapsing across variables (i.e., solving $Q = (1/n)\mathbf{Z}\mathbf{Z}'$) we obtain a $400 \times 400$ matrix of the correlations between people. If our data were time series measures of a number of stimuli, a type of data which is quite common in research in economics and political science, we could calculate either the correlations among the stimuli over time or the correlations between points in time over stimuli. Either approach might be appropriate for a particular research problem.

The data cube shows that there are six different ways in which we can compute correlations from a set of data that fill the cube (have several levels on each dimension). These are

| | |
|---|---|
| *R* correlation | Correlations between variables, summing across people at one point in time. |
| *Q* correlation | Correlations between people, summing across variables at one point in time. |
| *T* correlation | Correlations between times, summing across people for one variable. |
| *S* correlation | Correlations between people, summing across times for one variable. |
| *P* correlation | Correlations between variables, summing across times for one person. |
| *O* correlation | Correlations between times, summing across variables for one person. |

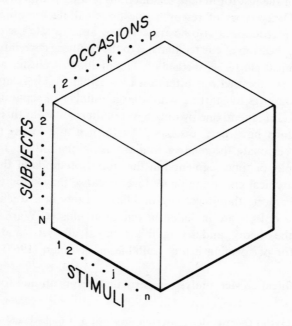

*Figure 9-1   The data cube. Each entry in the cube is the response of a particular subject to a particular stimulus on a particular occasion.*

Some of these approaches are less likely to prove useful than others. In psychology and education most multivariate research is done in the $R$ correlation paradigm. The important thing is that each of these ways of reducing the data cube to a correlation matrix provides a legitimate method for attacking some research problems.

Probably the most commonly used correlation after $R$ correlation is $Q$ correlation. It is often useful to know the degree of similarity in the patterns of responses made by individuals to several stimuli, and $Q$ correlation is one way of obtaining indices of similarity.* Once such a matrix of $Q$ correlations has been obtained, cluster analysis can be used to answer some interesting questions. For example, suppose that we have scores for 100 psychiatric patients, with various symptoms and diagnoses, on the 10 scales of the Minnesota Multiphasic Personality Inventory. By computing $Q$ correlations and performing a cluster analysis we could determine clusters of patients with similar MMPI profiles. By examining the symptoms and diagnoses of the individuals who cluster together on the basis of their test scores, we would obtain information about the usefulness of the test for making distinctions among people with different problems. Other types of research questions calling for other types of correlation matrices can also be answered by appropriate use of the data cube and application of cluster analysis to the resulting correlation matrix.

**Hierarchical clustering methods.** The method of cluster analysis on which we have focused our attention is one that builds relatively independent groupings by starting with a large number of separate observations and adding them, one by one. Several alternatives to this approach, some of which have been discussed by Borgen and Weiss (1971), are designed to evaluate the relative similarities among all the objects (variables, people, or time segments) in the study. Methods of this type are called hierarchical clustering procedures because they show the relative similarities among the objects by building clusters in a treelike fashion until all the objects are included in one large cluster. Ward (1963) has discussed the theory underlying this type of analysis, and computer programs for performing it are available in Veldman (1967) and Jones (1964).

Hierarchical cluster analysis is particularly appropriate for problems

---

* Alternatives to $Q$ correlation have been suggested by Cronbach and Gleser (1953), Mahalanobis (1936), and others.

such as those in biological taxonomy. Initially, each object (person, variable, or whatever) forms a cluster by itself. Then the two most similar clusters are joined, and a new value is computed for the cluster which is the average of the separate values for the members of the cluster. This is followed by either adding a third object to the cluster of two or by forming another cluster of two, depending on the relative similarities of the clusters. The two clusters with the smallest distance between them are combined to form a new cluster, and the value for that cluster is recomputed as the weighted average of its elements. The procedure goes on combining clusters until there is only one cluster, and it contains all the objects. Figure 9-2 illustrates how this procedure might work with a hypothetical set of data. First, elements 2 and 3 are combined to form a cluster, because they are more similar than any other pair, and a new point $[C_1 = (2 + 3)/2]$ is used to represent the cluster. Next, 7 and 8 are combined and represented by $C_2[= (7 + 8)/2]$. They are the next most similar. At the third stage element 1 is added to the $C_1$ cluster because it is more similar to the combination of these two variables than it is to the remaining single variables or the other cluster. The value representing this new cluster, $C_3$, is found by $C_3 = (2C_1 + 1)/3$, the weighted average of the $C_1$ cluster and the new element. The combining of clusters continues until all the clusters have been collapsed into one large cluster.

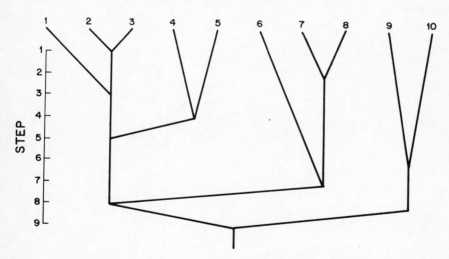

*Figure 9-2    Hypothetical example of a hierarchial clustering of 10 variables.*

One of the advantages of a hierarchical clustering is that it provides information about the relative similarities of the objects (e.g., variable 6 seems to be quite independent of all other variables, as do 9 and 10). However, it is often desirable to stop an analysis of this type before all variables have been combined into a single cluster, unless interest focuses on the order of cluster formation. If several clusters are desired, we have the problem, pointed out by Borgen and Weiss (1971), of when to stop. No satisfactory answer to this question has been given.

## SUMMARY

The class of procedures we have placed under the heading of cluster analysis is different in some important respects from other forms of internal factor analysis. Each variable is placed entirely in one cluster or another (or, perhaps, in no cluster) rather than having its variance split among several factors. Also, cluster analysis is free of some of the mathematical problems that can plague factor analysis. However, cluster analysis is not so mathematically elegant and rigorous as the factoring techniques we discuss in the remaining chapters. The decision to use cluster analysis must be made in light of the objectives of the particular study.

# 10

## THE GENERAL MODEL FOR FACTOR ANALYSIS

The remaining chapters in this book deal with the set of procedures that most often come to mind for investigators when the term factor analysis is used. This term, which has the weight of a long tradition behind it, is really a generic name for a variety of species of procedures. We have argued earlier that factor analysis is really a family level term in the taxonomy of statistical procedures and that the traditional term should be modified to *internal factor analysis*, reserving the term *external factor analysis* for procedures such as those discussed in Part II. However, because we will be dealing only with the former class and because the un-modified term factor analysis has traditionally been applied to these procedures, we use this term for simplicity and only modify it when necessary for clarity.

While we are dealing with overlooked semantic distinctions, it might be well to clarify the relationship between the word *component*, which we have been using, and the word *factor*, which we have generally avoided until now. When Hotelling (1933) published his derivation of a procedure for determining the composites of a set variables, he called those com-posites principal components. Later we shall see why he used the adjective,

**241**

but our concern now is with the noun. The term component refers to a composite of a set of variables developed from an ordinary correlation or covariance matrix. Thurstone (1931, 1947) proposed a method for analyzing a modified correlation matrix and gave the name multiple factor analysis to his procedure. Thurstone's modification involves using values less than 1.0 in the main diagonal of the correlation matrix. Although this distinction has been ignored in much of the literature dealing with composites of sets of variables, it seems to be a useful distinction to keep because it avoids confusion about the type of matrix that has been analyzed. The procedures discussed in Part II all involved unmodified correlation matrices. Thus the term canonical components refers to the composites developed in a canonical analysis. Canonical factors is a term that has sometimes been used to identify these composites, but since the term may lead to confusion between canonical analysis and a type of internal factor analysis developed by Rao (1955) which is called canonical factor analysis and because the composites resulting from a canonical analysis are really components in the sense that Hotelling used the term, it would seem better to call them components.

## THE FACTOR ANALYSIS PROBLEM

In very gross and oversimplified terms, the problem posed by factor analysis is to find a set of composites or latent variables of a single set of observed variables that will account for the variation among subjects on the observed variables. There are a variety of qualifiers that can be applied to dictate the character of the composites in a given situation. For example, we are usually interested in composites that are mutually independent, are as few in number as possible, and can account for the greatest possible proportion of observed variance. These qualifications of the solution and others will occupy our attention for the next two chapters.

**The dimensionality problem.** The best way to view a set of variables for factor analysis is to consider the variables as being plotted in a standardized space of people. We have seen in earlier chapters that the number of subjects ($N$) must be greater than the number of variables ($n$) so that there will not be linear dependence among the variables. Given $n$ linearly independent (but not uncorrelated) variables in a space of $N$ people, there is an *n*-dimensional subspace of the *N* space that completely contains the *n* variable vectors. The requirement of linear independence means that we must use *n* dimensions to completely describe this subspace. (The geometric relations among the variables are given by the correlation matrix. Because we are dealing with a standardized space, the correlations in the matrix are the cosines of the angles between the variable vectors. Therefore, analysis of the correlation matrix is an analysis of the variable space directly, and we can ignore the larger people space of *N* dimensions.)

If a complete description of the variable space is desired, *n* dimensions will be necessary, and the reader may ask why it is necessary to perform a complex statistical procedure to get as many composites as variables. Why not use the observed variables directly? One answer lies in the fact that the composites will be uncorrelated with one another.* The task of describing an *n* space in terms of *n* correlated dimensions is generally more difficult than describing the same space with *n* uncorrelated dimensions, because it is unnecessary to account for the relationships among the dimensions in the latter case.

A second and more telling reason why it is often useful to use uncorrelated composites rather than the original correlated variables is that the composites will generally be of unequal size and importance. The individual variables all have unit variance. Therefore, it is not possible to say that any one of them is more important or more pervasive in its influence than another. As we later see in detail, the composites will vary in size or amount of variance, making it possible to conclude that one is of greater importance as a dimension of variation than another. This feature of the composites is what has led to the distinction between factors and components and has been the source of much commentary on the proper approach to factor analysis.

Early in the history of factor analysis it was suggested that only the larger composites, those that accounted for a substantial proportion of the variation among subjects as reflected in the variation and covariation among variables, were of real scientific interest. This proposal is a logical consequence of theories about the nature of measurement in the behavioral sciences. Most of the research using factor analysis has been done by psychologists. The most prevalent theory of measurement in

---

* There are some procedures that generate correlated dimensions directly. With one special exception, we do not deal with these techniques. They are described in detail by Harman (1967).

psychology postulates that measurements involve a combination of true amount of the trait being measured and error of measurement. Although it may legitimately be argued that errors of some magnitude are present in measurements of all sorts, the errors present in psychological and other behavioral science measurement are generally considered to be relatively larger than those present in the hard sciences. The concern over measurement error and the resulting chance effects on the correlations among variables have caused psychologists to conclude that the smaller composites resulting from a factor analysis might reflect errors of measurement rather than true dimensions of variation among subjects. This has caused investigators to reject the smaller composites as real sources of variation and to focus attention on the major dimensions.

A second reason for rejecting the small composites may be referred to as parsimony of description. Factor theorists felt that the small composites accounted for such a small proportion of the variance of the set of variables that they did not contribute to understanding the relationships among the variables. In the interest of simplifying the description of these relationships, they argued, only the major sources of variation represented by the larger composites justified interpretation. The result of this reasoning was a rephrasing of the factor problem. The goal of factor analysis, according to this line of thought, is to describe and account for the relationships among a large number of observed variables using a relatively small number of uncorrelated latent variables. Thus Thurstone (1938) was able to account for a substantial proportion of the variance of 57 mental ability variables using nine major and four secondary factors.* The 13 latent variables provide a much more parsimonious description of the domain of abilities than do the 57 observed variables. It is sometimes argued that parsimony is the *raison d'etre* for factor analysis.

The question of what constitutes a major factor, of how large a factor must be to be of value, has been a problem for factor analysts for many years. A variety of answers to this question have been proposed, some of which are discussed in the next chapter, but none of them is completely satisfactory. The simplest answer is to avoid the problem entirely by retaining all the dimensions. This solution may be appropriate in a few special situations when parsimony is not an objective. For example, when

---

* Actually, Thurstone advocated allowing the factors to become correlated by rotation. See Chapter 11.

the accuracy of measurement is very high the small factors may reflect true rather than error variance. This might be the case for a problem in economics or medicine, but it would seldom be true of sociology or psychology. Also, there may be times when it is desirable to account for all the variance in a set of variables, regardless of the presence of error. However, in most cases the investigator will wish to use fewer composites than there are variables.

**The logic of reduced rank.** The reader will recall from our discussion of matrix terminology that the rank of a matrix is the number of linearly independent sources of variation among the variables in the matrix. We have seen that the rank of a correlation matrix is usually equal to the number of variables. Suppose we have a correlation matrix of three variables such that the variables can be represented graphically as shown in Figure 10-1, in which the plane of the page has been located to contain variables 2 and 3. Variable 1 projects up from the plane. We can view the

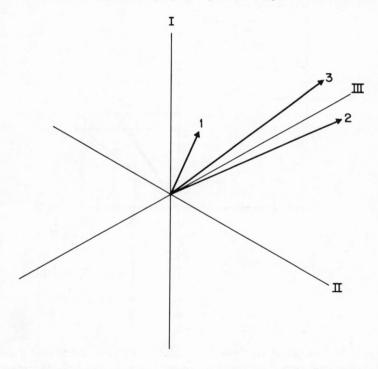

*Figure 10-1   Schematic vector representation of 3 variables in a 3-space.*

situation pictured in the figure as the true state of affairs. Using any three uncorrelated dimensions, we can completely describe this true state and can locate any variable vector exactly. Since the rank of such a matrix is 3, using three dimensions to describe it is called a full-rank solution. Such a solution is actually a model, in this case an exact one, of the relationships among the variables.

Using one composite to describe these variables also provides a model of the real world as it is represented by this particular set of variables. Such a model, which might look like that shown in Figure 10-2*a*, would be called a rank-one model. It represents a hypothesis that the relationships among the variables can be accounted for using only a single di-

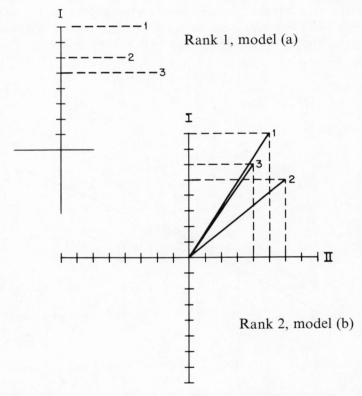

*Figure 10-2   Reduced rank models for the data in Figure 10-1. A rank 1 model (part a) and a rank 2 model (part b). Adding a dimension does not change the projection of a variable on an earlier dimension. The full rank model is presented in part c.*

mension. A different hypothesis would be that two dimensions are sufficient to account for the relationships. One such solution is depicted in Figure 10-2*b*. Both the one- and the two-dimensional models are called reduced rank models, because they are attempts to account for an *n*-dimensional situation by using fewer than *n* dimensions. Most applications of factor analysis represent attempts to fit a reduced rank model to a particular set of data. A variety of procedures are available to accomplish this result.

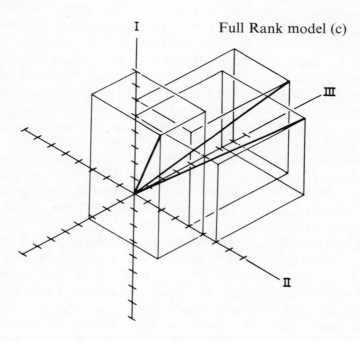

**THE GENERAL FACTOR ANALYSIS MODEL**

As we have used the term, factor analysis refers to a family of procedures used to develop latent composites of sets of observed variables under certain sets of restrictions. External factor analysis procedures are those in which the composites are found, subject to conditions outside the given set, generally conditions of maximum relationship to a variable, latent or observed, that is not considered part of the set. Internal factor analysis procedures are those that make use of information contained only within the set being analyzed. Several different types of restrictions could be placed on the solution, but they all involve nothing beyond the set itself.

**Factor analysis as multiple correlation.** One of the ways in which factor analysis can be viewed is as a particular variety of multiple correlation. Given a set of variables, we may ask what composite or composites of the set best predict the variables that are included in the set. If we take the composites one at a time, we might require that the first composite be determined in such a way that the average of its squared correlations with all of the variables in the set be a maximum. Next, we form a second composite under the restrictions that it be uncorrelated with the first composite and, given that condition, have maximum average squared correlations with the variables in the set. We may continue generating composites, each of which is uncorrelated with all previous composites and has maximum average squared correlations with the variables, until we have exhausted the rank of the correlation matrix.

The process of developing composites, whether under the above restrictions or some others, is generally referred to as the *extraction* of factors. The above set of restrictions results in a procedure called the *principal axis* method of extraction. We shall see shortly why the term principal axis is a good one.

Let us look at what we get from a principal axis solution. If we use a full rank model, we are able to describe completely the space occupied by the variables, but this would be true of any full rank model that analyzed a complete correlation matrix. However, the requirement that each composite (note that these are components when the main diagonal of the correlation matrix contains 1's, which is the case at present) have maximum average squared correlation with the variables in the set means that the first principal component will account for the maximum possible amount of variance of the variables. Each succeeding component will account for the maximum possible amount of the variance that remains after the previous components have been extracted. In most cases components beyond the first few account for relatively little variance.

Returning to the multiple correlation conceptualization of factor analysis, the reader should consider two points. First, the composites are analogous to the predictor variables. At any point in the extraction process the set of composites has been determined in such a way that the average of the squared multiple correlations of the set of composites for predicting the several observed variables is a maximum. The maximization of variance accounted for is a very important principle in the extraction processes of factor analysis. Without it there would be an infinite number of equally satisfactory solutions, making the value of any particular result a matter of personal preference. We shall see in the next chapter that a factor solution is sometimes improved by allowing personal preference to modify the initial factors, but preference is allowed only after extraction has been completed.

The second point, which the reader should carefully consider, is that the composites as extracted are orthogonal to one another. What we have is a set of uncorrelated, latent predictor variables for our multiple regression equation. In Chapter 6 we saw that the $\beta$ weight for a particular variable is a function of its correlation with the criterion and its correlations with the other predictors. But the composites, which are our predictors, are uncorrelated with each other. Therefore, for the case of orthogonal factors or components, the correlations of those factors or components with the individual variables *are* the $\beta$ weights for the predictors. These correlations are also the factor or component loadings in the sense in which that term was used in connection with external factor analysis.

There is one situation in which the factor $\beta$s and factor loadings are not equal. Some investigators, most notably Thurstone (1947) and Cattell (see Cattell and Jaspers, 1967) have argued that an initial factor solution should be transformed so that a set of correlated factors are obtained. We have more to say about this approach when we discuss rotation and simple structure, but the effect of such a procedure on the $\beta$ weights needs comment now.

The solution of the principal axis equations provides us with a matrix of $\beta$ weights for the factors. This matrix is called the *factor pattern matrix*, and its elements are called *pattern coefficients*. It is also possible to develop for any factor solution a matrix of the correlations between the factors and the observed variables. This is called the *factor structure matrix*, and its elements are *factor* (or component) *loadings*. When, as described above, the factors are orthogonal, the pattern coefficients are identical to the loadings. However, allowing the factors to become correlated by a transformation (the technical term is oblique rotation) destroys the isomorphism between pattern and structure. When factors are oblique, their locations are defined by the pattern coefficients, but they are described by the loadings.

Notice that there is no multiple correlation coefficient to be computed to indicate the accuracy with which the set of factors predict a given variable. Although the pattern coefficients are $\beta$ weights in the sense in which the term is used in multiple regression, they are weights to be applied to unmeasured variables. We do not have scores for individuals on the factors. In a later chapter we examine the problem of obtaining factor scores, but even when such scores are available, it is hard to imagine a purpose for which their multiple correlations with the variables might be used. (Actually, squared multiple correlations do enter the factor analysis model, but under the name of communality.)

**Geometric representation of factor analysis.** At several points in this book we have discussed the representation of a set of variables using vectors in a standardized space of people. Although it is almost certain that $n$ dimensions will be needed to completely describe $n$ variables, it is generally the case that the variables will be correlated with each other. When this happens the vectors representing the variables tend to fall more or less in groups. Those vectors that are closely grouped together represent variables that are highly correlated. Other groups of vectors may exist that are largely independent of (at right angles to) the first group, and still other groups may be at angles up to 180° from the first,

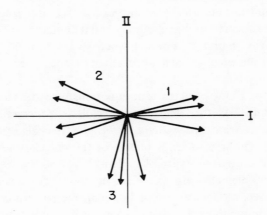

*Figure 10-3 Schematic representation of 10 variable vectors in 2 dimensions. Variables often form clusters around the factors.*

reflecting large negative correlations with the variables in the first group. Most factor analyses reveal the variables to be distributed more or less in this way, which is illustrated for 10 variables in two dimensions in Figure 10-3. The variables in each cluster are positively correlated among themselves; cluster 3 is essentially independent of clusters 1 and 2, and clusters 1 and 2 are highly negatively correlated with each other.

When plotted in this way, the variables tend to form a circular or elliptical figure. If each of $n$ variables has an exactly zero correlation with the others, the vectors will be the axes of a hypersphere of $n$ dimensions. As some of the variables are allowed to become correlated with others, the hypersphere becomes an $n$-dimensional hyperellipsoid. By discarding the shorter axes of the hyperellipsoid, the figure again approaches a hypersphere, but one of lower dimensionality. This is what happens when we apply a reduced rank model to a set of data in a factor analysis. We try to let a hypersphere of relatively low dimensionality do the work of the more complex ellipsoid. Figure 10-3 has been drawn so that a two-dimensional hypersphere can replace a ten-dimensional hyperellipsoid quite well. The eight discarded axes (which we were unable to represent anyway) were relatively short. The full rank model would look somewhat like a ten-dimensional pancake with only two major dimensions.

This description requires some qualification. So long as each variable has unit variance and there is no linear dependence among the variables, the $n$ points at the termini of the vectors define a hypersphere of $n$ dimensions. However, *the axes* of this hypersphere have different degrees of relationship with the variables. Those axes that are close to groups of variables will be highly related to the variables and account for large proportions of the variances of the variables. Axes that are not near groups of variables will have little variance associated with them. It is in this sense of proportions of variance projected onto the axes that the hypersphere becomes ellipsoidal. Figure 10-4 illustrates for three variables the argument being made. In the first part of the figure we see that the three variables do form a sphere, even though they are correlated. Variables 1 and 2 lie in a plane, and variable 3 is slightly above this plane. The second part of the figure shows the projections of the variables on axes I, II, and III, in that order. These are the principal axes of the sphere. Using the projections of the variables on these axes we obtain an ellipse that has I as its longest dimension, II as its next longest dimension, and III as its shortest dimension (I and II are in the plane). When we speak of ellipsoids and all variables having equal variance, it is the projections of the variables on the axes to which we are referring.

There is a geometric restriction on the factor extraction process which is the precise equivalent of the regression requirement that each succeeding factor have the highest possible average of squared correlations with the variables. A set of *n* correlated variables will form an *n*-dimensional

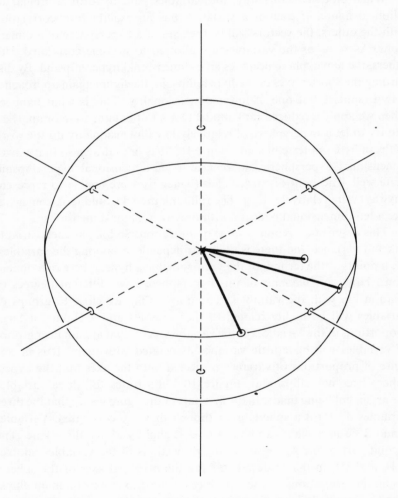

*Figure 10-4   Vector diagram of three variables. The variables themselves define a sphere (part a) while their loadings on the factors form an ellipse (part b).*

hyperellipsoid. The length of the hyperellipsoid in any direction from the origin is a function of the correlations among the variables. If we require that the first axis (factor or component) be located so that it is the longest possible axis of the figure, that axis will necessarily have the highest average of squares of correlations of the variables with the axis. By locating the second axis so that it is orthogonal to the first and follows the second longest dimension of the hyperellipsoid, we satisfy the same criterion for the remaining covariation. When each succeeding axis is placed at right angles to those preceeding it and through the longest remaining dimension of the figure, we obtain a set of principal axes that are precisely the same as those which result from the mean squared correlation criterion.

Figure 10-5 illustrates what happens when a principal axis solution is graphically presented for a simple problem. The five variables are represented by themselves in part *a* of the figure. No axes have been plotted, but it is easy to see that the variables tend to form a cluster. In part *b* the first axis has been inserted. The dotted lines represent the projections of the variables on the axis, and the average length of these

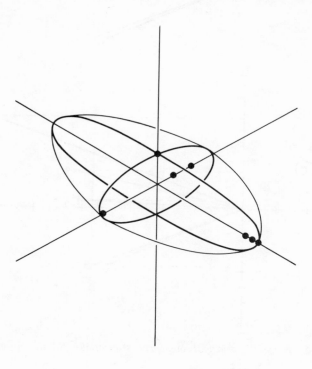

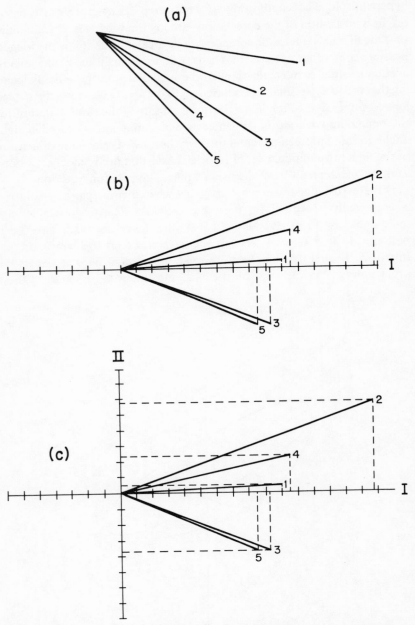

*Figure 10-5   Extracting principal components sequentially. See text for explanation.*

projections is greater than would be the case for any other axis. The second axis has been placed in part *c* of the figure so that it is orthogonal to the first axis and has the highest average projection. Part *d* shows the third axis placed to have maximum projections within the restrictions of orthogonality. There are two other principal axes that could be found, but the variables would have relatively small projections on them, perhaps representing measurement error. In most cases investigators find that *n*/3 or fewer axes are satisfactory for describing the relationships among the variables and that the first few axes account for most of the variance in the set.

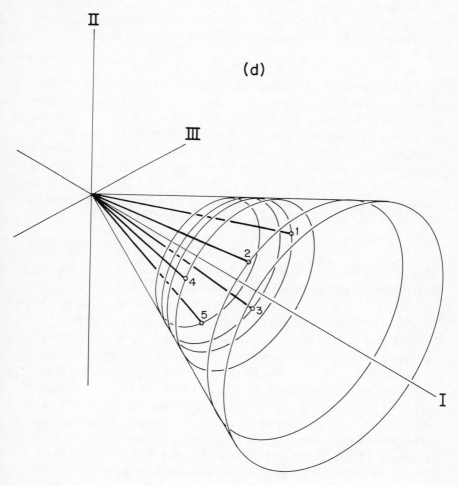

## THE MATHEMATICAL MODEL

The mathematical basis of factor analysis requires an assumption that the variance of a variable can be broken down into additive parts. The fundamental partition is one between that portion of its variance which a variable shares with other variables, its *common variance*, and that portion which is not shared, its *unique variance*. A full rank model, one which requires as many dimensions as there are variables in the set, postulates that all variance is common variance. Thus for $n$ variables there are $n$ common sources of variation, and the score of an individual on one of the variables can be expressed as the sum of his scores on the dimensions multiplied by the regression weights for those dimensions:

$$Z_{ij} = a_{j1}F_{1i} + a_{j2}F_{2i} + \cdots + a_{jk}F_{ki} + \cdots + a_{jn}F_{ni} \qquad (10.1)$$

The score of the $i$th individual on variable $j$ is the sum of the products of the regression weights of the $n$ composites for predicting the $j$th variable ($a_{jk}$) with the scores of the $i$th individual on each of these composites ($F_{ki}$). There are $n$ equations of this type that are necessary to specify the scores of one individual on $n$ variables. One advantage of a full rank principal component model is that the $n$ components will reproduce the scores of the individuals perfectly.

A more frequently used model for factor analyses postulates that the score of individual $i$ on variable $j$ can be expressed as

$$Z_{ji} = a_{j1}F_{1i} + a_{j2}f_{2i} + \cdots + a_{jm}F_{mi} + U_{ji} \qquad (m < n) \quad (10.2)$$

That is, the score is a combination of weighted scores on each of $m$ common sources of variations or factors, plus a *uniqueness* score for the person on that variable ($U_{ji}$). Stating the reduced rank model in this way and assuming that the $F$s represent latent variables that are normally distributed and standardized, we can see that each observed variable is composed of several common sources of variation and a unique source of variation. Because each of the $F$s has unit variance, the contribution of each factor to the variation in $Z_j$ is given by the square of its regression weight for predicting that variable (which is also the loading of the variable on that factor, $a_{jk}$). The sum of the squares of these factor loadings is the proportion of the variance of variable $j$ that is accounted for by the set of factors. This is also the proportion of variance that the variable has in common with the other variables in the set and is called its *communality*, which is symbolized as $h_j{}^2$,

$$h_j{}^2 = \sum_{k=1}^{m} a_{jk}^2 \tag{10.3}$$

The sum of squared factor loadings for any variable is its communality. Note also that the communality of a variable is the squared multiple correlation of that variable with the set of factors.

The uniqueness of a variable, which may be expressed as

$$u_j{}^2 = 1 - h_j{}^2 \tag{10.4}$$

can be broken down into two parts. It is assumed that each variable is measured with a certain amount of error and, because this error is random, error variance cannot contribute to common variance. If we symbolize error variance as $e_j{}^2$, we can express the reliability of a variable as

$$r_{jj} = 1 - e_j{}^2$$

However, it may be that not all of the reliable variance of a variable is common variance. We might have a variable that is quite reliable but unlike any of the other variables we have in our analysis. It would, therefore, have low communality. Reliable variance that is not shared with other variables is called specific variance or *specificity* ($s_j{}^2$). We may thus express the total variance of a variable as

$$Z_j{}^2 = 1 = h_j{}^2 + s_j{}^2 + e_j{}^2$$

and note that the reliability of a variable is

$$r_{jj} = h_j{}^2 + s_j{}^2$$

and its uniqueness s given by

$$u_j{}^2 = s_j{}^2 + e_j{}^2 = 1 - h_j{}^2$$

We may now restate the reduced rank factor analysis model in a more complete form as

$$Z_{ji} = a_{j1}F_{1i} + a_{j2}F_{2i} + \cdots + a_{jm}F_{mi} + s_{ji} + e_{ji} \tag{10.5}$$

to show that any score is composed of common portions, a specific portion, and an error portion. Many of the areas of controversy in factor analysis involve differences of opinion about how best to estimate these several components of the model.

The communality of a variable has an important place in the geometric view of factor analysis. Returning to our standardized space of people, we recall that each variable has a variance of 1.0, which is also the length of the variable vector. When we substitute any full rank set of reference axes for the $N$ dimensions of people, the lengths of the vectors are not changed, but their locations can be expressed as coordinates on the new set of axes. Their lengths are given by the square root of the sum of their squared projections on these axes which, for the full rank model, is unity. If we substitute a reduced rank model with $m$ $(< n)$ dimensions, the lengths of the vectors in the reduced space are changed, but they are still equal to the square root of the sum of squared projections of the vectors on the reduced set of axes. That is, the square root of the communality of a variable is the length of the vector representing that variable in the space of the common factors, and the result is the ellipse discussed above.

Another important feature of factor analysis which has both mathematical and geometric implications is the relationship between the factor loadings and the correlations among the variables. Again recalling our development of correlation, we find that the correlation between two variables is given by the mean of crossproducts of the coordinates of the two vectors in the people space. If we insert a full rank set of reference axes in place of people, the correlation between the two variables is given by the sum of crossproducts of the coordinates of the two vectors on the reference axes. A full rank model enables us to reproduce accurately the correlations among the variables from the component loadings. A reduced rank model permits us to reproduce the correlations among the variables more or less accurately, depending on the degree to which the variable vectors are completely contained within the space of the reference axes (i.e., have high communalities).

In the form of an equation the reproduced correlation between variables $X$ and $Y$ is given by

$$r_{XY}^* = \sum_{k=1}^{m} a_{Xk}a_{Yk} = a_{X1}a_{Y1} + a_{X2}a_{Y2} + \cdots + a_{Xm}a_{Ym} \quad (10.6)$$

which, in matrix form is

$$\mathbf{R}^* = \mathbf{A}\mathbf{A}' \quad (10.7)$$

where **R\*** is the matrix of reproduced correlations and **A** is the matrix of factor loadings (recall that we are dealing with orthogonal factors).

The **R\*** will differ from **R**, the original correlation matrix, by an amount that depends on the number and adequacy of the factors in the model. Any given reproduced correlation, $r^*$, between two variables differs from the original correlation between those variables, and this difference is called a *residual*, $\tilde{r}$, $(\tilde{r} = r - r^*)$.

$$\tilde{\mathbf{R}} = \mathbf{R} - \mathbf{AA}' = \mathbf{R} - \mathbf{R}^* \tag{10.8}$$

expresses these residuals in matrix form. As the number of factors extracted increases, the magnitude of the residual correlations in $\tilde{\mathbf{R}}$ decreases until, when $n = m$, all elements are zero and reproduction is perfect. The adequacy of the reproduced correlations as approximations to the originals (indicated by small residuals) was one of the first criteria used to determine when enough factors had been extracted.

## AN EXAMPLE

To illustrate the similarities and differences among the several varieties of internal factor analysis, we continue to use the Fleishman and Hempel (1954) data that were cluster analyzed in the last chapter. A full rank principal components analysis of the correlation matrix from Table 9-2 is shown in Table 10-1.

**Reading a factor table.** Table 10-1 is set up in the form in which the results of most factor analyses are presented. The first 15 rows of the table correspond to the 15 variables being analyzed, and the first 15 columns are the 15 principal components (full rank factors) that serve as reference axes for the variables. This part of the factor table is generally called the *factor matrix* [the matrix **A** of (10.7)]. Each of the entries in this 15 × 15 portion of the table is a factor loading. For example, variable 1 has a loading of .664 on factor 1, and variable 10 has a loading of − .420 on factor III. These factor loadings are both structure coefficients (i.e., correlations between the observed variables and the latent factor variables or composites) and pattern coefficients (regression weights for predicting the observed variables from the factors). We may, therefore, say that variable

**TABLE 10-1**

*Complete Principal Components Analysis of Data from Table 9-1.*

| Variables | 1 | 2 | 3 | 4 | 5 | 6 | 7 | 8 |
|---|---|---|---|---|---|---|---|---|
| 1 | 0.664 | 0.338 | 0.054 | -0.051 | 0.023 | -0.035 | 0.178 | -0.088 |
| 2 | 0.614 | 0.231 | 0.326 | -0.347 | -0.093 | -0.095 | 0.034 | -0.375 |
| 3 | 0.576 | 0.401 | 0.117 | 0.273 | 0.138 | 0.347 | -0.242 | -0.249 |
| 4 | 0.646 | 0.397 | 0.167 | 0.050 | 0.012 | 0.097 | -0.121 | 0.308 |
| 5 | 0.649 | 0.298 | 0.076 | 0.322 | -0.145 | 0.013 | 0.163 | 0.187 |
| 6 | 0.717 | 0.125 | -0.110 | 0.049 | -0.069 | -0.275 | -0.013 | 0.074 |
| 7 | 0.691 | 0.040 | 0.013 | -0.244 | 0.029 | -0.392 | -0.168 | 0.019 |
| 8 | 0.526 | 0.044 | -0.565 | 0.292 | 0.217 | -0.283 | 0.252 | -0.133 |
| 9 | 0.658 | -0.155 | -0.019 | -0.470 | 0.174 | 0.019 | 0.032 | 0.261 |
| 10 | 0.559 | 0.010 | -0.420 | -0.279 | -0.154 | 0.507 | 0.252 | 0.053 |
| 11 | 0.615 | -0.346 | -0.252 | 0.040 | 0.098 | 0.110 | -0.378 | -0.200 |
| 12 | 0.678 | -0.381 | -0.105 | 0.007 | -0.113 | 0.041 | -0.153 | -0.101 |
| 13 | 0.500 | -0.452 | 0.289 | 0.151 | -0.425 | -0.006 | 0.286 | -0.173 |
| 14 | 0.447 | -0.394 | 0.375 | 0.107 | 0.165 | 0.101 | 0.242 | 0.016 |
| 15 | 0.615 | -0.353 | 0.111 | 0.229 | -0.180 | -0.009 | -0.180 | 0.300 |
| Eigenvalue | 5.670 | 1.368 | 0.975 | 0.839 | 0.770 | 0.731 | 0.629 | 0.605 |
| % of variance | 37.8 | 9.1 | 6.5 | 5.6 | 5.1 | 4.9 | 4.2 | 4.0 |

*(Continued)*

| Variables | 9 | 10 | 11 | 12 | 13 | 14 | 15 | Communality $\Sigma a^2 = h^2$ |
|---|---|---|---|---|---|---|---|---|
| 1 | 0.072 | 0.390 | -0.396 | -0.237 | 0.151 | -0.003 | -0.067 | 1.00 |
| 2 | -0.225 | 0.083 | 0.108 | 0.214 | -0.139 | -0.106 | 0.197 | 1.00 |
| 3 | -0.160 | -0.173 | 0.112 | -0.030 | 0.217 | 0.085 | -0.187 | 1.00 |
| 4 | 0.118 | -0.297 | -0.263 | 0.031 | -0.115 | 0.057 | 0.295 | 1.00 |
| 5 | 0.241 | 0.195 | 0.198 | 0.295 | -0.176 | -0.087 | -0.192 | 1.00 |
| 6 | -0.107 | -0.168 | 0.183 | -0.421 | -0.118 | -0.319 | -0.052 | 1.00 |
| 7 | 0.240 | 0.006 | 0.221 | -0.060 | 0.092 | 0.398 | -0.033 | 1.00 |
| 8 | -0.158 | -0.112 | -0.072 | 0.205 | 0.082 | 0.061 | 0.118 | 1.00 |
| 9 | -0.082 | -0.148 | -0.090 | 0.228 | 0.215 | -0.186 | -0.229 | 1.00 |
| 10 | 0.008 | 0.041 | 0.199 | -0.129 | -0.041 | 0.122 | 0.109 | 1.00 |
| 11 | 0.363 | 0.123 | -0.017 | 0.058 | 0.044 | -0.246 | 0.138 | 1.00 |
| 12 | -0.155 | -0.039 | -0.256 | 0.007 | -0.404 | 0.172 | -0.226 | 1.00 |
| 13 | 0.172 | -0.268 | -0.062 | -0.019 | 0.203 | 0.003 | 0.008 | 1.00 |
| 14 | 0.036 | 0.032 | -0.103 | -0.113 | -0.122 | 0.038 | 0.056 | 1.00 |
| 15 | -0.367 | 0.284 | 0.072 | 0.015 | 0.175 | 0.048 | 0.153 | 1.00 |
| Eigenvalue | .576 | .550 | .509 | .499 | .463 | .436 | .380 | 15.00 |
| % of variance | 3.8 | 3.7 | 3.4 | 3.3 | 3.1 | 2.9 | 2.5 | |

1 has a correlation of .664 with the first factor or that .664 is the least squares standard regression weight ($\beta$ weight) for factor I for predicting scores on variable 1 from scores on the set of factors. Both these interpretations hold for all 225 factor loadings in the table; however, attention usually centers on the structure coefficients. It is these values, considered as structure coefficients, that are used for "interpretation" of the factors.

There are two rows and a column of the table we have not yet discussed but which provide important information for understanding the results of a factor analysis. The row labeled eigenvalue contains the sums of the squares of the factor loadings in each column. The sum of squares of the loadings of the variables on the first factor is 5.670. Because each of the factor loadings is a correlation and the square of a correlation is the amount of variance in one variable which is accounted for by the other variable, the eigenvalue can be interpreted as the variance of the factor. Thus the variance of the first factor is 5.670, and the variances of the remaining factors are similarly given by their eigenvalues.

The last row of the table, which is labeled % of variance, is the eigenvalue for that column (factor) divided by the sum of the variances of all of the variables. In the present full rank example each variable has a variance of 1.0; thus the sum of these variances is 15. When we begin to deal with reduced rank models in the next chapter we will need to change our thinking about this row of the table, but, for the present, we may consider the percent of variance to be the eigenvalue divided by the number of variables, for example,

$$\frac{5.670}{15} = .378 = 37.8\%$$

An important feature to notice when considering the eigenvalues and percentages of variance is that they become progressively smaller for each succeeding factor. The first factor has the largest eigenvalue, the second factor has the next largest eigenvalue, and each additional factor has a smaller eigenvalue. This is not surprising when we recall the restrictions placed on the factors when they were extracted. We required that the first factor account for the largest possible amount of variance in the set and that each succeeding factor account for the maximum possible variance remaining after previous factors had been removed. This means that the early factors are generally the most important, in that they account for more of the variance in the set of variables than do later factors.

The last column of a factor table is also important for interpreting the results of a factor analysis. This column, labeled communality or $h^2$, contains the sums of squared factor loadings for each of the variables (the sum of squares across each row of the factor matrix). As we have seen, these communalities give the amount of variance of each variable that is accounted for by the set of factors. In a full rank model the set of factors accounts for all the variance of each variable. Thus all the communalities are equal to 1.0. Because each of the variables has unit variance, this means that all the variance of each variable is common variance. Each of the variable vectors is completely contained within the space of the factors.

The value in the lower-right-hand corner of the table (15.0) is the total amount of variance accounted for by all the factors. It is also the total amount of variance of all the variables for which the factors account. Thus it is both the sum of the communalities and the sum of the eigen-values. As such, it is an index of the total amount of common variance in the set of variables as described by the factors. This value divided by the number of variables gives the proportion of variance in the total set of variables, which is common variance. For our full rank model this is 100%.

**Reproducing the correlation.**   We use the first three variables to il-lustrate the procedure by which the correlations among the variables can be reproduced from the factor loadings. If we multiply the loading of variable 1 on each factor by the loading of variable 2 on the same factor and sum these values, the result is .450127. The last three digits of this number are due to random error generated by the computer, since the value read in was .450. Applying the same procedure to rows 1 and 3 yields a value of .409071, which is also off slightly, because of random errors. Taking the sum of crossproducts in the same way for rows 2 and 3 gives a reproduced correlation of .380359. The reader is invited to assure him-self that the correlations can be reproduced accurately from the factors by multiplying any pair of rows and checking the resulting value against the correlations in Table 9-1.

There is one other way to take sums of crossproducts in the factor matrix: down pairs of columns. When this is done, the resulting values are the correlations between pairs of factors. But, we already know what these values should be! Recall that we placed a restriction on our second and succeeding factors that each account for the maximum possible amount of that variance that is left over after the earlier factors have been removed. We also said that the second factor was to be the longest axis of the subspace that is independent of (or orthogonal to) the first axis. Thus we expect the correlation between any two factors to be zero. Finding the correlation between factors I and II by taking the sum of crossproducts down the columns, we obtain a value of .00794. This value is so close to zero that we may reasonably claim to have obtained orthogonal factors. Any other pair of factors from Table 10-1 can be shown to be uncorrelated (except for random error) by applying the same procedure.

**Reduced rank model.** There are 14 reduced rank models we could choose to describe our set of 15 variables. We must take the factors in order of magnitude, because the definition of one of them, say the fifth, depends on the ones that have gone before it. The fifth factor is defined in part by the fact that it is independent of the first four. Therefore, our model of rank $k$ must include the first $k$ factors. Given this condition and others to be discussed in the next chapter, we will use a model of rank 2. Using the first two factors also makes it possible to describe the factors graphically, a procedure which has frequently been used to aid in factor interpretation.

The first two factors of Table 10-1 have been isolated in Table 10-2. The loadings for these factors are precisely the same as those of the full rank model, which is what we would expect, given the method by which the factors were found. The eigenvalues of these factors are, of course, also the same as those in the full rank model; however, the last column of the table, which contains the communalities, is different. These values are all substantially less than 1.0 because the two-dimensional space defined by the factors does not completely contain any of the variables. The communalities give the proportion of variance of each of the variables which is contained in the space of the two factors. The communalities and eigenvalues are computed as sums of squares of rows and columns, respectively, as outlined earlier.

There is another difference between the two factor tables; Table 10-2 has an additional row. The row labeled % of total contains the values called % of variance in Table 10-1. Because the total variance in the original correlation matrix remains the same regardless of the number of factors, the values in this row are still the eigenvalues divided by the number of variables. The last value in this row indicates that the two factors taken together account for 46.9% of the total variance in the set of variables (7.038/15). The new row, labeled % of common indicates the relative amount of common variance (the total variance of all retained factors) that is contributed by each of the individual factors. For factor I this is $5.67/7.038 = 80.6$, and for factor II it is $1.368/7.038 = 19.4$. The sum of the variances of all factors retained in the model is the total common variance.

The factor loadings of the 15 variables on the first two factors are presented graphically in Figure 10-6. The fan-shaped array of variable vectors which we see in the figure is fairly typical for plots of variables on the first two factors of almost any factor analysis of variables that are all positively correlated, as is the case for our data. However, the plot shows us that there are a few fairly tight groups within the fan. Variables 1 to 5 seem to form one group which have moderate positive loadings on both factors. Variables 6, 7, 8, 9, and 10 form a group that has loadings

**TABLE 10-2**

*First Two Principal Components from Table 10-1.*

| | Factor | | |
|---|---|---|---|
| *Variable* | I | II | $h^2$ |
| 1 | .664 | .338 | .555 |
| 2 | .614 | .231 | .430 |
| 3 | .576 | .401 | .493 |
| 4 | .646 | .397 | .575 |
| 5 | .649 | .298 | .510 |
| 6 | .717 | .125 | .530 |
| 7 | .791 | .040 | .627 |
| 8 | .526 | .044 | .279 |
| 9 | .658 | −.155 | .457 |
| 10 | .559 | .010 | .313 |
| 11 | .615 | −.346 | .498 |
| 12 | .678 | −.381 | .605 |
| 13 | .500 | −.452 | .454 |
| 14 | .447 | −.394 | .351 |
| 15 | .615 | −.353 | .503 |
| Eigenvalue | 5.670 | 1.368 | 7.038 |
| % of total | 37.8 | 9.1 | 46.9 |
| % of common | 80.6 | 19.4 | |

almost exclusively on the first factor, and variables 11 to 15 have positive loadings on the first factor and negative loadings on the second.

Factor plots were much more common 20 to 30 years ago than they are today. The major reason for this is that factor plots were used to derive alternative sets of factors from an initial set such as we have obtained. The alternative solutions were found by changing the locations of the factors in the plot to achieve better description of the factors. This procedure is known as *rotation* and is discussed in more detail in the next chapter. Using plots such as that in Figure 10-5 the factor analyst would reposition the factors to pass through groups of variables, hoping to thus simplify his description. Each pair of factors would be plotted in turn and the locations of the factors adjusted pairwise until the investigator was satisfied with the result. For a factor analysis resulting in $k$ factors, there would be $[k(k-1)]/2$ factor plots to be inspected and adjusted. There are

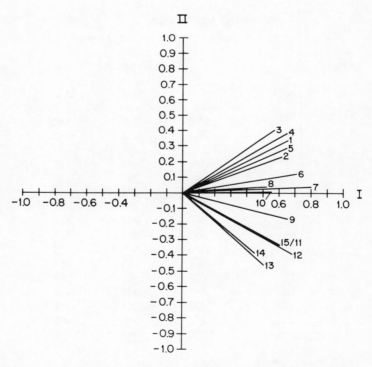

*Figure 10-6    Plot of the first two principal components from Table X-2.*

now several objective rotation methods that can be performed by computer. Consequently, graphic rotation by inspecting factor plots is relatively rare now.

It was said earlier that each factor solution in a sense represents a model of the observed correlations. We have seen that a full rank model is essentially a complete model of the correlation matrix in that it provides, within a small random error margin, perfect reproduction of the correlations. Let us now examine some reduced models to see what happens to our ability to reproduce the observed correlations.

The most severely reduced model possible is a single factor model such as we had in Figure 10-2. Such a model for our present example would yield a reproduced correlation of .408 for variables 1 and 2 (.664 × .614 = .408), which is in error by only .042. However, the single factor model would yield the same correlation between 1 and 15, this time in error by .118. If we were to reproduce the entire correlation matrix, we would find that many of the residual correlations (differences between the original and reproduced correlations) would be quite large. We can improve the adequacy of our reproduction by taking additional factors. For example, for the two-factor model in Table 10-2 we obtain .486 as the reproduced correlation between the first two variables and .289 as the reproduced correlation between variables 1 and 15. In the second case this is quite an improvement. The addition of more factors will gradually improve the situation until we reach the limiting case of 15 factors and virtually perfect reproduction of the correlations. One problem that has confronted factor analysts almost from the beginning, and which we discuss in the next chapter, is how many factors are enough, either for adequate description of the variables or for reproduction of the correlations.

## SUMMARY

In this chapter we have discussed conventional factor analysis in its most general form. We have seen that factor analysis is a procedure for finding composites of a set of variables subject to the restriction that the first composite account for the largest possible amount of variance in the set and that each succeeding composite do likewise, subject to the additional restriction that each composite in turn be uncorrelated with all those which have gone before. It is almost always the case that $n$ factors will be required to account completely for the variance found in a correlation matrix of $n$ variables. The use of fewer than $n$ factors generally requires that some of the information in the correlation matrix be discarded.

A factor analysis following the principles we have discussed results in a factor matrix that contains coefficients called factor loadings. So long as the factors are required to be independent of each other, these factor coefficients can be interpreted as either structure coefficients or pattern coefficients. The structure coefficients are correlations of the ordinary (product moment) sort between the observed variables and the latent factor variables. Pattern coefficients are standard regression weights for predicting the observed variables from the factors. The two types of coefficients are equivalent for orthogonal factors.

We have also seen that the variance of a variable can be broken down into a common factor portion (its communality) and a unique factor portion (its uniqueness). Unique variance may be further divided into specificity and error of measurement. The reliability of a variable is the sum of its common variance and its specific variance. In addition, factors can be viewed geometrically as the axes of a hypersphere defined by the variables. The sum of squared loadings of the variables on these axes are called the eigenvalues of the factors, and the sum of squares of the loadings of a variable on all of the axes is its communality. The square root of the communality of a variable is the length of its vector in the space defined by the factors.

# 11

# MODIFICATION OF THE
# FACTOR ANALYSIS MODEL

The factor analysis model presented in the preceding chapter was based on what is called an *unreduced correlation matrix*. This means that the main diagonal of the matrix contains the variances of the variables (remember that they are in standard score form), and the off-diagonal elements are the correlations among the variables. The introduction of reduced rank models early in the development of factor analysis lead Thurstone and others to question the use of unreduced correlation matrices. The reasoning was that if only common variance was being extracted from the matrix, then only common variance should be included in the matrix. This has resulted in a variety of strategies by which one makes some estimate of the common variance of each variable and uses these estimates in place of 1.0 as the elements of the main diagonal.

In this chapter we discuss several of the procedures that have been proposed for reducing the correlation matrix. However, many of these procedures depend heavily on the rank chosen for the factor model. The two problems, number of factors and reduction of the correlation matrix (also known as estimating the communalities of the variables), are so closely related that they are difficult to separate. Insofar as possible we

**269**

first discuss choice of rank, then communality estimation. Next, the problem of maximizing the interpretability of the factor solution is discussed, and finally we consider some alternate proposals for extracting the factors.

## CRITERIA FOR FACTOR SOLUTIONS

Before we begin our discussion of the practical problems faced by the user of factor analysis we must set the stage by considering the criteria that factor theorists have used to define the adequacy of a factor analysis.

**Reproduction of the correlations.**    The first and foremost criterion of adequacy for a set of factors is the accuracy with which they reproduce the correlations from which they were developed. It has even been said (Cattell, 1958) that this is the only completely justifiable standard that can be applied. However, this test can only be used to compare solutions that use the same number of factors, because we will amost always find that the adequacy of reproduction of the correlations is better for larger numbers of factors. In the extreme and perhaps trivial case, $n$ principal components will reproduce perfectly (except for rounding errors) the correlations among $n$ variables. However, this approach does provide a method by which different factor solutions for the same data may be compared.

A potentially useful way to judge whether an additional factor adds enough information beyond that already contained in a set of factors is to compare the matrix of residual correlations obtained with the extra factor to the matrix obtained without it. Because the residuals are the differences between the original and the reproduced correlations, the addition of a useful factor to the set should result in a noticeable decrease in the size of these residuals. Thurstone (1947) and others have used the criterion of insignificant residuals as an index the adequacy of a given set of factors, and McNemar (1942) has discussed the theory of residual distributions in an effort to provide a statistical test of when the residuals are sufficiently small. There has been little effort to compare different factor solutions on this criterion.

**Minimum Rank.**    A second criterion for the adequacy of a factor solution, one closely tied to the reproduction of the correlations, is the criterion of minimum rank. In his early writings on factor analysis

Thurstone proposed that the diagonal values of the correlation matrix be changed from unities to those that would yield the best reproduction of the reduced correlation matrix with the fewest factors. (Notice that the communality problem is rearing its head.) The rank of a reproduced correlation matrix is always exactly equal to the number of factors used to compute it. Thurstone argued that the best factor solution was the one requiring the smallest number of factors (minimum rank of the reproduced correlation matrix) to approximate adequately the original set of correlations.

Although Thurstone's proposal has some logical merit, it has been severely criticized on mathematical grounds. Guttman (1954) has shown that the minimum rank of a matrix that exactly reproduces the original correlations is seldom less than $n - 1$ and that there are usually many different sets of $n - 1$ factors which will provide this reproduction. The principal components solution described in the preceding chapter provides a unique solution of rank $n$ for an unreduced matrix. When an analysis of a reduced matrix is desired and a model of rank $n - 1$ is used, it is generally not possible to find a unique best set of factors. Thus we are again left with the problem of deciding what is adequate reproduction of the correlations. We may choose at will from a variety of rank $n - 1$ solutions, or we may use some model that does not perfectly reproduce the correlations and then struggle with the problem of insignificant residuals. In either case, we have not gotten very far.

Wrigley (1958, 1959) has listed three objections to the use of minimum rank as a criterion for the adequacy of a set of factors.

1. There is no way to determine minimum rank.
2. The approximations that have been used generally yield more factors than most investigators will accept.
3. The procedure makes no allowance for sampling error.

His conclusion is that this approach is unworkable as a means for determining either the number of factors or the communalities of the variables, which are the two problems on which discussions of the adequacy of factor solutions focus.

There is another problem we must confront when we attempt to use minimum rank as a logical base for a reduced rank factor model. If we analyze an unreduced correlation matrix the number of factors will be too high, thus we would like to use a reduced matrix of some kind. This reduction generally involves inserting some estimate of the variable communalities in the main diagonal of the correlation matrix. But, we defined the communality of a variable as the sum of its squared loadings on the factors. This is the crux of the problem. Our decision about optimal reduced rank is based in part on estimates of common variance, and our estimates of the communalities depend in part on the number of factors we choose to retain. For the most commonly used approaches to factor analysis these two problems cannot be separated. The most frequently used solution to the problem is to make an arbitrary decision about one feature, either rank or communality, and to obtain the solution to the other by direct computation. A variety of arbitrary decisions have been proposed for both rank and communality, which we now examine.

## THE NUMBER OF FACTORS PROBLEM

There are two basic types of criteria which may be employed to decide how many factors are sufficient, those based on logical considerations and those based on statistical theory. Also, it is possible to identify two schools of thought on the general problem. The first school, which we might call the conservatives and which includes mainly British factor theorists, recommends keeping as few factors as possible. A statement by Burt (1966, p. 287) is representative of this position.

> In factor analysis it is an axiom of the method that we refrain from postulating any further factors unless it is clearly shown that the factor or factors already extracted fail to account (within the limits of statistical significance) for the entire amount of variance presented by our data.

Some American theorists take a position at the opposite extreme and might, by contrast, be called liberals. For example, Cattell (1958, p. 801) has argued that "the number of real common factors operating in any set of variables is indefinitely large and can be infinite." Proponents of this position would argue that each variable is composed of a large number of sources of variance that are shared with some other variables somewhere in the universe. Because of the restriction that only a relatively small number of these variables are included in any given study (or, indeed, may be known to science) only a limited number of the common factors that compose the variable can appear. If the selection of the other variables in the study had been different, others of the common factors that compose the variable might have been found. This position has a certain amount of merit. However, it is a rather uncomfortable position for a supposedly deterministic science to be in. What it boils down to, to paraphrase Tom Lehrer,* is that factor analysis is like a sewer; what you get out of it depends on what you put into it. The factors that emerge from any factor analysis are a function of the variables selected for study. A particular selection will precondition certain factors to appear in the analysis and prohibit others from appearing.

There are practical limits to the number of variables that can be handled in a single study. If an investigator is interested in determining the internal structure of a particular set of variables, then he need not be particularly concerned with the implications of Cattell's statement. His variables constitute a population for him, and he may wish to analyze them into as few common factors as are necessary. If, however, the investigator wishes to determine the composition of his variables within the context of a larger population of variables, he must exercise great care in the selection of the variables to be included in a particular factor analysis. We have more to say on this topic later. In the present context it is sufficient that the reader be aware that his decision about the number of factors to retain in a particular study must take account of the purpose of the study. Choice of a criterion for the number of factors must be based in part on a consideration of purpose.

A large number of criteria have been proposed for the number of factors to extract in factor analysis. These may be conveniently divided into two groups, those based on statistical considerations and those based on logical or psychometric considerations. Some of the most prominent and potentially useful criteria in each group are discussed below.

**Psychometric criteria.** The most widely known and used criterion for the number of factors is what is known as the Kaiser criterion (Kaiser, 1960). This approach is based on a principal components analysis (unreduced correlation matrix) and advocates retaining only those factors that have eigenvalues greater than 1.0. This means that for a factor to be retained, it must account for at least as much variance as does a single variable, a requirement that has substantial intuitive appeal.

---

* "An Evening Wasted with Tom Lehrer," Tom Lehrer Records, 1959.

Other arguments have been advanced for using the Kaiser criterion. In his initial statement, Kaiser noted that "for a principal component to have positive Kuder–Richardson reliability, it is necessary and sufficient that the associated eigenvalue be greater than one . . ." (1960, p. 145). This logic was extended by Kaiser and Caffrey (1965) to the domain sampling of variables concept described above. They argued that if the variables in a study can be considered to be a sample from a universe of variables, the resulting factors must then also be a sample from a universe of factors. Under these assumptions, only those factors that satisfy the Kaiser criterion can be considered to be generalizable to the universe.

The Kaiser criterion is actually the application of an algebraic derivation by Guttman (1954). Guttman was concerned with the minimum number of factors that would be required to account for the off-diagonal elements of a correlation matrix. He showed that the number of eigenvalues of the matrix that were greater than 1.0 constitutes a "weak lower bound" on the number of factors. His derivation indicates that the Kaiser criterion yields the smallest number of factors that can possibly do the job. The number of factors necessary cannot be less than the number indicated by the Kaiser criterion, and is probably more.

In the same paper Guttman showed, further, that a second, "stronger" lower bound and a third "universally strongest" lower bound on the number of factors exist. The second, lower bound is the number of eigenvalues of the reduced correlation matrix (with the square of the highest off-diagonal correlation for that variable inserted in the main diagonal) that are nonnegative ($\geq 0$). He called this value the nonnegative index. The third or strongest Guttman lower bound for the number of factors is the nonnegative index of the reduced correlation matrix with the squared multiple correlation of each variable predicted from all of the other variables used in the main diagonal as an estimate of that variable's communality. Guttman showed that the rank of the correlation matrix, and hence its proper number of factors in the sense of Thurstone's minimal rank notions, will always be at least as great as this strongest lower bound.

There are some practical reasons why Guttman's two stronger lower bounds have not been widely applied. The second lower bound is in the position of a middle child, generally ignored because attention tends to focus on the extremes. There is little evidence available on how well this estimate of the number of factors would actually work in practical situations. The strongest lower bound will almost always yield more than $n/2$ factors for $n$ variables. A criterion this liberal has two drawbacks. First, the later factors to be extracted will generally have small loadings and, therefore, be hard to interpret. Second, it is difficult to obtain good estimates of the communalities when such a large number of factors have been retained. We have more to say about this problem when we discuss communality estimation.

It is important to keep some points in mind when considering use of Guttman's strongest lower bound. First, it is a bound, not an exact value. Although it is an algebraic lower bound, it is probably best used as a practical upper limit on the number of factors that can be kept when a reduced correlation matrix is analyzed. The reason for this is that factors kept beyond this limit are, in one sense, imaginary. Recall that an eigenvalue is equal to the sum of squared factor loadings of the variables on that factor. Factors beyond Guttman's lower bound, when squared multiple correlations are used in the main diagonal, have negative eigenvalues. For the sum of squared factor loadings to be negative, at least some of the factor loadings must be imaginary numbers. Even though Kaiser (1961) has shown that Guttman's strongest lower bound is valid for all proper communalities (those between 0 and 1), it would seem that it constitutes a practical upper bound.

Harman (1967) has proposed a modification of Guttman's strongest strongest lower bound which takes into account the fact that there will be negative eigenvalues. When a reduced correlation matrix (values less than unity, generally communality estimates in the main diagonal) is factored completely ($n$ factors are obtained) the sum of the eigenvalues will be equal to the sum of the communalities. Because some of the eigenvalues will be negative, the sum of the positive eigenvalues as obtained using Guttman's strongest lower bound will be greater than the sum of the communalities (but always less than $n$). This means that in a sense one is extracting more than 100% of the common variance. Harman has suggested that factors be extracted until the sum of the eigenvalues is equal to the sum of the communalities. Although the results of such a decision rule have not been studied, it has logical appeal and it has the advantage that it can be applied in any situation where prior estimates of the communalities are available.

The Kaiser criterion or Guttman's weak lower bound has been the center of a substantial amount of controversy among factor analysts.

For many it has considerable appeal because it seems to offer a complete program for factor analysis. Once one has decided to analyze an unreduced correlation matrix and retain only those factors with eigenvalues greater than unity, most of the problems are over. This approach has such widespread appeal that most computer centers offer a packaged option for obtaining this type of analysis. However, a substantial number of factor analysts believe that the easiest way may not be the best way.

Use of the Kaiser criterion has been attacked on both theoretical and empirical grounds. On the theoretical side, Cliff and Hamburger (1967) have pointed out that Guttman's assumptions in deriving all three of his bounds, namely, that the variables under study are sampled from a population and that the individuals constitute a population, also apply to Kaiser's derivations. They note that it is almost impossible to justify these assumptions for an empirical investigation. Citing an empirical study, Humphreys has argued that "It is quite clear that, when $N$ is large [his sample contained over 8000 individuals], factor extraction can proceed far beyond the criterion suggested by Kaiser and results obtained that are apparently real and dependable" (1964, p. 465). Of course, Humphreys is arguing about the statistical significance of the factors, but his criticism seems justified. Cattell feels that the Kaiser criterion is useful, but only for a restricted range for the number of variables. He has claimed that "the Kaiser test . . . cuts off too soon when variables are few ($n \leq 20$) and too late when they are many ($n > 50$)" (Cattell, 1966b, p. 207). Viewing factor analysis in terms of weighted linear regression, Butler and Hook (1966) come to a somewhat different conclusion. They claim that "If attention is centered upon common factors determined solely by the side [off-diagonal] elements of the correlation matrix, the number of latent roots [eigenvalues] of the correlation matrix above unity must constitute its maximum number of factors" (p. 563).

The foregoing is an example of some of the controversy that has surrounded the Kaiser criterion. Obviously, the decision rule is the one which we applied to get the reduced rank model for our example in Chapter 10. We have more to say about the Kaiser criterion after we have looked at problems of communality estimation and rotation, but now we must turn our attention to other suggestions for deciding on the number of factors.

Our next psychometric criterion for the number of factors to retain comes from Cattell (1966a, 1966b). After reviewing and discarding many of the common criteria (1966b), Cattell proposed (1966a) a test that appears to be very simple and easy to use. He reasoned that the sizes of the eigenvalues would cease to change very much after the nontrivial common variance had been removed from a correlation matrix. The plot of these eigenvalues should show a steady decrease until all of the non-trivial variance has been removed and then flatten out into a "scree." In Figure 11-1 we have plotted the eigenvalues for our factor matrix in Table 10-1. The points describe a fairly nice curve.

Figure 11-1 illustrates the basic problem with the scree test. Cattell recommends that one look for "breaks" in the curve, but we have no adequate definition of what a break is. Also, the appearance of the curve is heavily dependent on the scales of the two axes. An investigator who wishes to find fewer factors need only expand the horizontal axis to have the scree start earlier. Of course, expanding the vertical axis will yield a later scree and more factors.

Actually, the scree test is the most recent in a long line of "curve-break" or "root-staring" methods for determining the number of factors. Early in

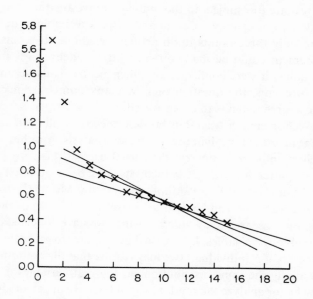

*Figure 11-1   Plot of eigenvalues from Table 10-1. Straight lines indicate various possible "screes."*

the history of factor analysis, McNemar (1942) raised a basic objection to all such criteria, that some other criterion is necessary to determine objectively when the curve has broken. The three lines drawn through the curve in Figure 11-1 illustrate three different interpretations of a curve break. One would suggest two factors, the second, three, and the third line, which, given our scale for plotting, looks most reasonable, would require us to retain six factors. Cattell has attempted to answer this criticism of indeterminancy by suggesting that the scree test can be objectified by "taking the first differential of the curve and finding at what point it departs significantly from zero" (Cattell and Jaspers, 1967, p. 41). However, even disregarding the fact that the scale on the horizontal axis is ordinal, it is likely that the equation for the curve of the eigenvalues will be complex and difficult to determine, which cancels one of the advantages claimed for the test, its simplicity.

When used as a quick approximation or for descriptive purposes, the test may be useful, particularly if it is used in conjunction with and for confirmation of other tests, such as one of Guttman's bounds or one of the statistical procedures described below. But curve-break tests such as the scree test are too subject to the individual investigator's judgment to be used alone. For example, Cattell and Jaspers feel able to determine a scree from only three points in an artificial example. It seems unlikely that a judgment could be made on such little evidence in a study with real data unless it were confirmed by other means. Thus, in addition to its other problems, the question of how many points are necessary to determine a scree remains to be answered.

The psychometric criteria thus far described have all been ones which the average investigator might use in a factor analysis. Another approach, one which is unlikely to interest the casual user at this time but which deserves the attention of factor theorists, has been outlined by Horn (1965) and Linn (1968). These authors suggest that Monte Carlo methods with random variables can be used to answer the question of the number of factors. Suppose we have a study in which we have collected data from $N$ individuals on $n$ variables. Let us use the computer to generate $n$ random scores for each of $N$ individuals (we may require that the random variables have essentially the same distributions as the real variables if we wish). If we now factor analyze the random variables, we can get an idea of how large an error factor we might get in our real data. Horn suggests retaining only those factors from the real data that have eigenvalues exceeding the largest eigenvalue obtained from the random variables. This procedure will ensure retaining no error factors, but, unless the random data correlation matrix is a reduced one, the random factors approach will yield even fewer factors than the Kaiser criterion. Therefore, it is subject to the same criticisms of conservatism.

An interesting sidelight of the Horn and Linn studies is that both authors report the occurrence of curve breaks in the real data factors at about the point where error factors should begin. This finding may lend some support to the scree test.

It is fairly obvious that the random-factors approach requires more work at the present time than the informational gain from its use merits. Too little is known about the behavior of random variation and its effects on factor analysis. However, it would be possible to include a random variable generator in factor analysis programs. The cost in terms of running time would probably not be excessive on the larger, faster computers. Thus the investigator who is particularly concerned to avoid any possibility of an error factor coming out of his analysis could feel reasonably safe. It is possible that this approach may become generally available and the method of choice once we know more about it.

**Statistical criteria.** The eigenvalues and factor loadings that are computed from a set of data are statistics in the same sense that the mean and standard deviation are. Therefore, they have sampling distributions that, although complex and not precisely known, can be used to determine the approximate statistical significance associated with a particular solution. Those criteria for the number of factors that we cover in this section depend in one way or another on distributions of statistics.

We have seen that the factors can be used to reproduce the original correlations and that the accuracy of this reproduction seems to be related to the number of factors used. When more factors are extracted, the reproduced correlations more closely approximate the original correlations and the residuals become smaller. In the extreme case in which there are as many factors as variables, the residuals are zero except for rounding error. Several theorists have proposed that the decision of when enough factors have been extracted should be based on the distribution of these residual correlations.

One of the early advocates of using residual correlations to aid in the decision on the best number of factors to extract was Thurstone. His method of centroid factor analysis (see Thurstone, 1938 or 1947 for a discussion and bibliography) required the use of residuals at almost every stage of the extraction process. After the first factor was extracted, the residuals were computed, and the second factor was extracted from a modified residual matrix. Succeeding factors were extracted from successive residual matrices. Because Thurstone's procedure was cyclic, extracting one factor from a matrix, computing residuals, extracting another factor, computing new residuals, and so on, and because the computations required considerable labor, it was very desirable to avoid extracting unnecessary factors. Thurstone recommended that factoring cease when the residual correlations were small, because it was unlikely that there would be any meaningful factors left in the data.

Several investigators proposed ways in which one might decide when the residuals were too small to justify further factoring. This work was reviewed, summarized, and found unsatisfactory by McNemar (1942), who proposed a refinement. McNemar suggested that an adequate criterion must be based on sample size (to account for sampling error in the number of factors) and the number of variables. Working from this statistical position, he derived an adjusted formula for the standard error of a residual which was based on the standard error of a partial correlation. Under this proposal, factors would be extracted from the data until the distribution of residuals had a standard deviation no greater than the standard error of a zero-order correlation. Although this approach would seem to merit consideration, it has not been widely used. See Sokol (1959) for a discussion of the residuals approach to the number of factors problem.

A second statistical direction of attack on the problem of the number of factors has been the evaluation of the eigenvalues for statistical significance. One of the first approaches of this type was developed by Lawley (1943). His basic notion was that because there are $n$ eigenvalues for a matrix of $n$ variables, a null hypothesis is that all of the eigenvalues are equal. If this is the case, there can be no common factors, because the only time when the eigenvalues can all be equal is when the variables are uncorrelated. That is, for the eigenvalues all to be exactly equal, the correlation matrix must be an identity matrix. Lawley derived a function of the eigenvalues which has a $\chi^2$ distribution (approximately). If one or more of the eigenvalues are significantly larger than the others, a statistically significant $\chi^2$ will result; otherwise, the $\chi^2$ will be nonsignificant.

Lawley's test was derived under a restrictive set of assumptions.

Bartlett (1950) showed that insufficient degrees of freedom would make the test inappropriate, but he found it to give good results for large samples. In its place, Bartlett proposed a $\chi^2$ test for the equality of all the eigenvalues, one which is less sensitive to small sample sizes. Bartlett's test, which Cooley and Lohnes (1971) call the sphericity test because it tests the hypothesis that the variable vectors form the independent axes of a hypersphere, is given by

$$\chi^2 = -\left[(N-1) - \frac{2n+5}{6}\right]\log_e |\mathbf{R}| \tag{11.1}$$

where   $N$ = sample size,

   $n$ = number of variables, and

   $|\mathbf{R}|$ = the determinant of the correlation matrix

The $\chi^2$ statistic has $[n(n-1)]/2$ degrees of freedom. Cooley and Lohnes (1971) cite a study by Knapp and Swoyer that indicates that Bartlett's test gives useful results.

As originally derived, Bartlett's test applied to the equality of the whole set of eigenvalues. Lawley (1956) has generalized Bartlett's test to the situation in which one is testing the equality of the $n - k$ eigenvalues that remain after the first $k$ factors have been extracted. Each factor extracted before the $\chi^2$ dips below the chosen level of significance can be considered statistically significant. The modified test is given by

$$\chi^2 = -\left[(N-1) - \left(\frac{2n+5}{6}\right) - \frac{2k}{3}\right]\log_e D_{n-k} \tag{11.2}$$

where
$$D_{n-k} = \frac{|\mathbf{R}|}{\left[\prod_{j=1}^{k} \lambda_j \left(\frac{n - \sum_{j=1}^{k} \lambda_j}{n-k}\right)^{n-k}\right]}$$

$\chi^2$ has $\dfrac{(n-k)(n-k-1)}{2}$ degrees of freedom, and

$\lambda_j$ = the eigenvalue of the $j$th already-extracted factor, and

$k$ = the number of factors that have been extracted.

The modified Bartlett test outlined here may be used fruitfully at any point in the factoring process. The computer program for factor analysis given by Cooley and Lohnes routinely computes this test for every factor extracted. Others, such as Burt (1952), Maxwell (1959), and Tobias and Carlson (1969), have also suggested the routine use of Bartlett's test in conjunction with a factor analysis. Because the test is easily computed and interpreted, its systematic use should be encouraged. However, sample size has a direct and substantial effect on the test. There is an increasing amount of evidence that statistical significance is not a sufficient condition on which to base a decision about importance (e.g., Hays, 1963). To be worthwhile, a factor should be statistically significant *and meaningful*. This implies that the judgment must be based on both logical and statistical criteria. Therefore, statistical significance by a test such as Bartlett's may be considered a necessary condition for a factor to be of value, but it is not a sufficient condition. To merit interpretation and further study a factor should be both meaningful and real (real in the sense that we may be reasonably certain that it is not due to sampling error).

A variety of other statistical tests have been proposed for use in factor analysis. However, they have generally been derived to accompany particular methods of extracting factors. For this reason, further consideration of statistical tests is deferred until we have discussed the factoring methods with which they are associated. However, we shall see that there is no single solution to the problem of the number of factors to use. It is up to each investigator to consider his data in light of several indicators and arrive at a rational (and personal) decision.

## THE COMMUNALITY PROBLEM

The equation for the reduced rank general factor model, which we discussed in Chapter 10, includes a uniqueness term that refers to non-common variance. In almost every case in which a reduced rank factor model is applied to a set of data we find that some of the variance of each variable is unique to that variable. As we have also seen, unique variance can be divided into a portion that is reliable but specific in the sense that it is unrelated to other variables which have been included in the set, and a portion that is identified as error of measurement variance. This notion, that the diagonal elements of the correlation matrix might be composed in part of error variance and in part of specific variance, has been the major source of the communality problem. The problem is this: what can one put in the diagonal of the correlation matrix that will eliminate unique variance? Answer, the communality. How do I determine the communality? To this second question there are almost as many answers as there are factor theorists. We shall discuss briefly some of the proposed solutions, with the full realization that none of them provides a definitive answer to the problem.

As is the case with many of the facets of factor analysis, discussions of the problem of communality estimation generally begin with the work of L. L. Thurstone. Early in his work Thurstone was aware of the need to use something other than unities in the main diagonal of the correlation matrix if one were to analyze only the common variance of the variables. Because it was not possible to "know" the communalities before performing the analysis, Thurstone recommended using each variable's highest correlation with the other variables in the set as an estimate of what the communality might be. With this highest off-diagonal element for each variable inserted in the main diagonal as an estimate of the communality, the factor analysis, using centroid extraction, could proceed.

Thurstone would have been the first to acknowledge that using the highest off-diagonal element to estimate the communalities provided a rough approximation at best. It is important in evaluating his work to remember that most of his developments and writings were in the B.C. (before computers) era. The science of computing has developed so rapidly in the years since his death that it is hard for us to appreciate the amount of labor involved in a factor analysis. Therefore, although the procedures Thurstone developed have largely fallen out of use, the importance of his pioneering work is not diminished by the fact that he made some concessions to accuracy out of pure practical necessity. Even some of the procedures we might now take to represent minimal methodology would have resulted in an intollerable increase in labor 25 years ago.

Another method of estimating communalities was proposed by Thomson (1934). As a member of the British school of factor theorists, headed by Spearman, Thomson believed in a single general factor of human ability and several specific factors. He recommended that the first principal component of the unreduced correlation matrix be found and that the square of the factor loading for each variable on this general factor be used as an initial estimate of its communality. By inserting these estimates in the main diagonal of the correlation matrix, a reduced matrix could be obtained. This reduced matrix could then be factored and new estimates of the communalities obtained from the squared factor loadings on the new first factor. Thomson recommended that this procedure be repeated until the communality estimates did not change from one factoring to the next.

The procedure suggested by Thomson has come to be known as iteration for the communalities. It may readily be generalized to the case where more than one factor is extracted; however, its use requires some prior decision about the number of factors. Some authors (e.g., Cattell, 1958) have suggested that iteration for the communalities is highly desirable in any study involving factor analysis. When sufficient access to a computer is available, this is certainly true for most factor problems. However, it must be remembered that the suggestion to iterate was made in the context of a single factor model. When the number of variables becomes even reasonably large, say, 15 to 20, and more than a very few factors are retained, the labor of an interative approach soon comes to exceed its value unless a computer is available. It is for this reason that American factor theorists sought other ways to estimate the communities.

We have already discussed Thurstone's alternative to iteration. A second method, one which is generally more sound and is still frequently used, was suggested by Roff (1935). He reasoned that the square of the multiple correlation of each variable in the set with all of the other variables in the set should provide a good estimate of the communality of each variable. This approach has considerable intuitive appeal because the squared multiple correlations (often call SMCs) do indicate the proportion of variance of each variable that is predictable from the other variables in the set. However, the labor involved in calculating SMCs for large sets of variables kept all but the most hardy investigators from using this approach for about 20 years.

Advocates of the use of SMCs as estimates of the communalities received a double-barreled boost in the 1950s from developments in computers and from the work of Guttman (1956). Computers eased the computational burden and Guttman showed that *"If the Spearman–Thurstone hypothesis is correct for a given* **R**, *then the $\rho_j^2$ [SMCs] must almost always be very good approximations to the $h_j^2$ when n* [the number of variables] *is large.* (Conversely, if the approximation is bad for many $\rho_j^2$, then the Spearman–Thurstone hypothesis of a limited number of common factors must be false)" (p. 274, italics in original). Guttman was also able to show that the SMCs become better approximations to the communalities of the variables as the number of variables increases. Wrigley (1957, 1958, 1959) has agreed with Gutmann in both these conclusions.

An even stronger defense of the use of SMCs has been made by Butler (1968). He argued that the correlation matrix consists of descriptive statistics (correlations) and that the SMCs may therefore be considered to be observed communalities rather than estimates. However, Butler goes on to note that this position is based on assumptions that are not met in practice.

Up to this point our discussion has focused primarily on lower bound estimates of the communalities. Both the SMCs and the highest off-diagonal elements provide estimates of the minimum values for the communalities. It is often desirable to determine an upper bound as well as a lower bound to yield a range within which the communalities may be expected to fall. One upper bound is, of course, unity. However, a much better limit is available. Recall from our discussion in the previous chapter that the variance of a variable can be broken down into a common portion, a specific portion, and an error portion. By definition, the error variance can never become common variance. But, the definition of common variance for a particular variable depends on what other variables are included in the study. What is common variance when the variable being considered is included in one set of variables may be specific variance when a different set is chosen. For example, a test of mechanical aptitude may have very little variance in common with a set of general intelligence measures, but the same instrument might have high communality when the set is composed of other mechanical ability tests. It could be argued on grounds such as these that the distinction between common and specific variance should not be made until after the factors have been extracted. If we wish to use an estimate of the common *plus* specific variance in the diagonal of the correlation matrix, then each variable's *reliability* is the value we should use. The reliability of a variable provides an upper bound for its communality.

Mulaik (1966), in recommending the use of reliabilities as communality estimates, has argued that factor analysis is a technique for psychometric inference. That is, factor analysis is seen as a technique for discovering the underlying (latent) common dimensions or sources of variation in the universe of possible measures. The communality of a variable in this infinite universe of variables is "the reliability of the variable (except possibly for an infinitesmally small discrepancy)" (p. 123). "The distinction between common and specific reliable variance may not be tenable in a universe of variables" (p. 124).

It might seem at first glance that the reliabilities of the variables offer a definitive solution to the communality problem if we are willing to accept the assumption of a potentially infinite universe of variables. However, there is a second and less obvious assumption that must be made and which is generally unacceptable, namely, that *the* reliabilities of the variables are known. Several authors, most notably Cronbach *et al.* (1972), have demonstrated that there is no single reliability for any variable. Observed reliabilities depend on a host of factors, which include the characteristics of the sample, experimental conditions, and type of statistics (e.g., split half, test-retest), used in the reliability study. In fact, reliabilities are subject to at least as much frustration and uncertainty as other communality estimates are.

Before the reader throws up his hands in hopeless confusion and despair, we should add that precise values for the communalities, although desirable, are not essential to obtaining useful results from a factor analysis. Good estimates will serve for most purposes, and these can be obtained fairly painlessly by a general procedure we outline shortly. If one is willing to make certain assumptions about his data, there are also some procedures in which the communalities are defined by a particular mathematical formulation. These are discussed as special factoring models later in this chapter.

At this point the reader is painfully aware that there is no definitive solution to the communality problem. Each of the initial estimates we have discussed has several drawbacks. In fact, the concept of common variance is inextricably bound to the selection of variables for study, so that we may rightfully conclude that we must choose one of two assumptions if we wish to use the notion of common variance. On the one hand, we may choose to assume that the variables of our study constitute a population and that communality is defined by that population. On the other hand, we may assume that our variables are a sample from some universe and that defining reliability defines communality. Under this second assumption the problem becomes one of psychometric method rather than statistical method, and we will not consider it further. The first assumption keeps the problem a statistical one, but we must digress to consider another problem before suggesting a solution.

## THE ROTATION PROBLEM

When we first drew the distinction between internal and external factor analysis we said that the composites were defined under certain restrictions. For internal factor analysis the primary conditions were that each factor account for the greatest possible amount of variance in the set and that each factor be statistically independent of all preceding factors. Under these restrictions the basic factor model presents no particular problems. Starting with an ordinary correlation matrix (unreduced), there is a defined mathematical solution for internal factor analysis, just as there is for each of the varieties of external factor analysis. However, as we have seen, factor theorists have not been content to let things stand in this nice, uncomplicated way. The desire for parsimonious description has led to the number of factors problem, and an aversion for unique variance has created the communality problem. To complicate matters still further, factor theorists have, in general, been unsatisfied with the patterns of factor loadings provided by principal axis extraction and have spent large amounts of time trying to improve on the mathematically defined solutions provided directly by extraction. The justification for most of this activity is that it clarifies the meaning of the factors.

Once again, we may hold L. L. Thurstone chiefly responsible for creating or identifying the problem. In one of his early statements, *The Theory of Multiple Factors* (Thurstone, 1933), this illustrious troublemaker noted that any reference frame in the form of factors could serve for describing the relations among variables. Not only is it unnecessary that the chosen dimensions be principal axes,* they need not even be orthogonal. Thurstone felt that the primary criterion for the location of the axes was that they be interpretable as meaningful psychological or other substantive constructs. So convincing was his argument, and so useful his results, that few social scientists today would disagree with the principle. Thus in a few masterful strokes Thurstone succeeded in stripping factor analysis of all vestiges of objectivity, leaving his contemporaries and successors to do with their factors as each has seen fit.

Perhaps one of the reasons why factor analysis has developed in the particular and, for statistical methods, peculiar way that it has is because

---

* Thurstone actually extracted his factors by the centroid technique, because the labor of principal axis factoring for his large sets of variables would have been virtually impossible with the computing hardware then available.

its development has been largely at the hands of psychologists. Psychological data, riddled as they are with measurement and sampling error and being objectifications of covert variables such as intelligence and personality, are particularly susceptible to the sort of freelance interpretation which is possible through factor analysis. However, factor analysis has proved a very useful tool in organizing vast quantities of psychological data, and the chaos that might have resulted from Thurstone's opening of Pandora's box (Thurstone actually created his own set of boxes to support some of his notions) has been of significant benefit. With these introductory remarks, we turn our attention to the problem of selecting an optimum reference frame, the rotation problem.

**Logic of rotation.** Two basic characteristics which Thurstone described as necessary for a factor solution to be acceptable were psychological meaningfulness and simple structure. To require that a solution be meaningful within the domain under study seems similar to coming out in favor of motherhood and apple pie, and most writers on factor analysis have ignored the problem. However, a study by Armstrong and Soelberg (1968) showed that it is possible to obtain "meaningful" results from an analysis of *random data!* This would lead one to conclude that the fact that a solution makes sense is no guarantee that substantive scientific constructs have been identified. Indeed, when venturing into a new domain of inquiry, it may prove difficult to find "meaningful" factors until a great deal more is learned about the nature of the domain. It would appear that the criterion of meaningfulness is, ultimately, a necessary condition for an adequate factor solution, but it is far from sufficient, unless the meaning derives from a program of research.

It is Thurstone's second criterion of acceptability that has seen by far the greater activity on the part of factor theorists. The search for simple structure has led to considerable debate, headscratching, and progress. As originally proposed, the criterion of simple structure required that each variable have a high loading on few factors (preferably only one) and that for any given pair of factors, there be few variables with high loadings on both factors in the pair. If we represent high loadings by $x$s and low loadings by 0s, an ideal simple structure might have a form such as that shown in Figure 11-2. In this case each variable has a high loading on a single factor. The advantage to a solution such as this is that there is no doubt about which variables are contributing variance to which factors. The task of interpretation is eased if the solution approximates simple structure.

Simple structure solutions do not normally pop out of correlation matrices in the process of factor extraction. As we saw in Chapter 10, the most common result of extraction is a single general factor, on which almost all variables have moderate to high loadings, and a set of bipolar factors. Because the variables with which Thurstone worked were generally positively correlated, the occurrence of bipolar factors (which were not considered meaningful) and the complex first factor led him to seek some transformation of the factors. If we recall that the factors represent one arbitrary set of reference axes from an infinite number of such sets, we see that the rotation problem involves finding another arbitrary set of axes that more adequately fulfills our needs. For example, in Figure 10-5 we had a set of 15 variables which were represented in a space of two factors, one general and one bipolar. In Figure 11-3 we have the same variables plotted, but two other possible sets of reference axes have been included. Note that the positions of the variables remain constant; it is the axes that are being moved. Also, the lengths of the variable vectors remain unchanged by the process of rotation. This means that the communalities of the variables must remain constant, thus placing the limiting condition on the factor loadings that the sum of squared loadings for each variable be unchanged by the rotation process.

Factors

|   | 1 | 2 | 3 | 4 |
|---|---|---|---|---|
| 1 | 0 | $x$ | 0 | 0 |
| 2 | 0 | 0 | $x$ | 0 |
| 3 | $x$ | 0 | 0 | 0 |
| 4 | 0 | $x$ | 0 | 0 |
| 5 | $x$ | 0 | 0 | 0 |
| 6 | $x$ | 0 | 0 | 0 |
| 7 | 0 | 0 | $x$ | 0 |
| 8 | 0 | 0 | $x$ | 0 |
| 9 | 0 | 0 | 0 | $x$ |
| 10 | 0 | $x$ | 0 | 0 |
| ⋮ | ⋮ | ⋮ | ⋮ | ⋮ |
| $n$ | 0 | 0 | 0 | $x$ |

*Figure 11-2   Schematic factor matrix showing ideal simple structure. X's represent non-zero loadings.*

**Graphic rotation.** In the early days of factor analysis factors were interpreted by inspection of plots such as the one in Figure 11-3. However, in most cases there would be more than two factors being considered for interpretation. Because any given plot could handle only two factors at a time, it was necessary to make a separate plot for each possible pair of factors and to rotate each pair independently. Then new factor loadings would be calculated and plots of the rotated factors prepared for further inspection. If additional rotations were considered necessary, the whole process would be repeated, one pair of factors at a time.

The mathematics of rotation, although requring the use of matrices, is really quite simple. What is involved in going from the pair of factors I, II in Figure 11-3 to the pair I′, II′ is the movement (rotation) of the axes through an angle of some number (call it $\theta$) of degrees. In order to compute the new factor loadings for I′ and II′ it is necessary to find a transformation matrix (**T**) that will reflect the rotation of these factors through an angle of $\theta$ degrees. The new matrix of rotated factor loadings **B**) may then be found by multiplying the original factor matrix (**A**) by the transformation matrix.

$$\mathbf{B} = \mathbf{AT} \qquad (11.3)$$

The problem at this point, of course, is to find a satisfactory transformation matrix. Recall that we decided to consider only situations in which the factors were orthogonal to each other. (Oblique rotations or allowing the factors to become correlated with each other are favored by many factor analysts. However, these solutions are conceptually and mathematically much more difficult. Good discussions of oblique solutions are found in more advanced texts, such as Harman (1967) and Thurstone (1947). The orthogonality requirement means that as each pair of factors is rotated the relations between the factors must be retained at 90°. It can be shown (see Harman, 1967, Chapter 12) that a transformation matrix of the form

$$\mathbf{T} = \begin{bmatrix} \cos\theta & \sin\theta \\ -\sin\theta & \cos\theta \end{bmatrix}$$

will do the job for two factors. Given the angle of rotation and a set of sine and cosine tables, it is very easy to compute the loadings of the variables on the new factors.

Graphic rotation becomes somewhat more complex when we consider more than two factors. Because we are limited to two factors at a time in factor plots, the number of plots needed increases rapidly. For $m$ factors, $[m(m-1)/2]$ separate plots are required. Also, because our transformation matrix must be of order $m \times m$, it will also have to be changed from that given above. We use the first three factors from Table 10-1 (reproduced in Table 11-1) to illustrate the general principles of graphic rotation.

The plot of the first two factors from Table 11-1 is given by axes I and II in Figure 11-3. The plots of factor pairs I − III and II − III are shown in Figures 11-4a and b, respectively. In each of these figures the pair of factors being represented lies in the plane of the page and the other factor is perpendicular to that plane. In the full space of three dimensions the

**TABLE 11-1**

*Loadings of 15 Variables on the First 3 Components from Table 10-1.*

| | Factors | | | |
|:---:|:---:|:---:|:---:|:---:|
| *Variable* | I | II | III | $h^2$ |
| 1 | .664 | .338 | .054 | .558 |
| 2 | .614 | .231 | .326 | .537 |
| 3 | .576 | .401 | .117 | .506 |
| 4 | .646 | .397 | .167 | .603 |
| 5 | .649 | .298 | .076 | .516 |
| 6 | .717 | .125 | −.110 | .542 |
| 7 | .691 | .040 | .013 | .479 |
| 8 | .526 | .044 | −.565 | .600 |
| 9 | .658 | −.155 | −.019 | .457 |
| 10 | .559 | .010 | −.420 | .489 |
| 11 | .615 | −.346 | −.252 | .561 |
| 12 | .678 | −.381 | −.105 | .616 |
| 13 | .500 | −.452 | .289 | .538 |
| 14 | .447 | −.394 | .375 | .496 |
| 15 | .615 | −.353 | .111 | .515 |
| Eigenvalue | 5.670 | 1.368 | .975 | 8.013 |
| % of total | 37.8 | 9.1 | 6.5 | 53.4 |
| % of common | 70.8 | 17.1 | 12.1 | |

locations of the variables are fixed. They appear to change position because our perspective on the space changes. Each plot represents the projection of the variable vectors onto a different plane. These three different perspectives are illustrated in Figure 11-5. When viewing the plot of factor pair I-II, it is as though we were at the end of the III axis (labeled perspective I-II) looking at the plane I-II. The I-III plot (Figure 11-4a) gives the picture we would get looking down from the end of the II axis (perspective I-III) onto plane I-III, and Figure 11-4b shows what we

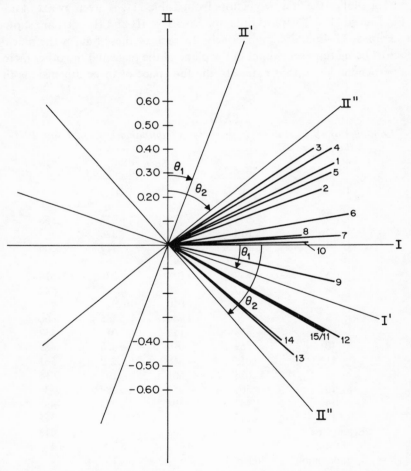

*Figure 11-3   Two possible rotations ($\theta_1$ and $\theta_2$) of the components from Figure 10-5. Note that the components move while the variables retain their positions.*

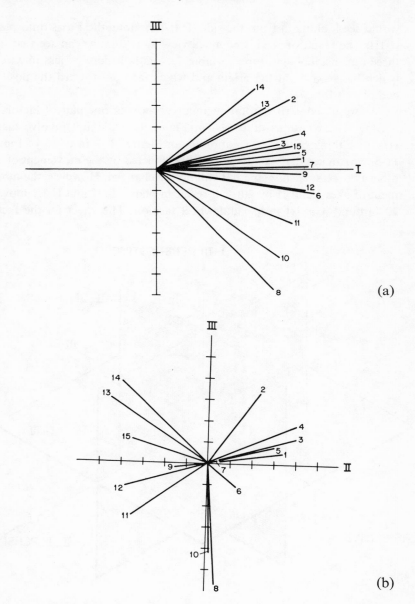

*Figure 11-4    Plots of 15 manual dexterity variables on factors I and III (part a) and II and III (part b) from Table 11-1. Figures 10-5 and 11-4 (a and b) provide a complete plot of these 3 factors.*

would see looking in from the side (P II-III) along the I axis onto plane
II-III. The result of all this complex viewing is that we can see that in a
three-dimensional space the variable vectors fall more or less in a cone
which has its apex at the origin and which fans out toward the positive
end of axis I.

As we noted earlier, rotation involves moving one pair of factors at
a time. The rotations described in Figure 11-3 did not involve factor
III. If we think of the 20° rotation ($\theta_1$ in Figure 11-3) in terms of Figure
11-5, we can see that axis III will not be affected by the movement of the
other two axes. Also, plane I-II will not be changed. However, the movement of axes I and II by 20° will result in planes I-III and II-III moving
20° around axis III in a paddlewheel fashion. The effect on the factor

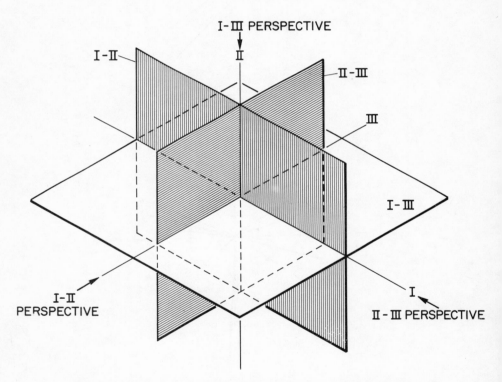

*Figure 11-5   Three planes showing the different perspectives presented in Figures
10-5 and 11-4 (a and b). Figure 10-5 is the I–II perspective, Figure 11-4(a) is the I–III
perspective, and Figure 11-4(b) is the II–III perspective.*

loadings of this rotation of factors I and II 20° in a clockwise direction is shown in Table 11-2. The transformation matrix by which the factor loading matrix is multiplied is

$$\mathbf{T} = \begin{bmatrix} \cos 20° & \sin 20° \\ -\sin 20° & \cos 20° \end{bmatrix} = \begin{bmatrix} .9397 & .3420 \\ -.3420 & .9397 \end{bmatrix}$$

Inspecting the three factor plots, we may decide that additional rotations are necessary. The plot of I-III suggests that a clockwise rotation of about 50° will be necessary to clear up the bipolarity of factor III. (Note that Figure 11-4a does not include the effect of our first rotation. Strictly speaking, the factors should be replotted after each rotation.) Such a rotation involves a **T** matrix of the form

$$\begin{bmatrix} .6428 & .7660 \\ -.7660 & .6428 \end{bmatrix}$$

To make a long story short, an additional rotation of 30°, this time in a counterclockwise direction, of factors II and III gives results that

**TABLE 11-2**

*New Factor Loadings for Factors I and II After a Rotation of
20 Degrees ($\theta_1$ in Figure 11-3).*

|     | I     | II    |
| --- | ----- | ----- |
| 1   | .508  | .545  |
| 2   | .498  | .427  |
| 3   | .404  | .574  |
| 4   | .471  | .594  |
| 5   | .508  | .502  |
| 6   | .631  | .363  |
| 7   | .636  | .274  |
| 8   | .479  | .211  |
| 9   | .671  | .079  |
| 10  | .522  | .201  |
| 11  | .696  | −.115 |
| 12  | .767  | −.126 |
| 13  | .624  | −.254 |
| 14  | .555  | −.217 |
| 15  | .699  | −.121 |

look satisfactory. The **T**-matrix has the form

$$\begin{bmatrix} .866 & -.500 \\ .500 & .866 \end{bmatrix}$$

Notice that for a counterclockwise rotation the location of the "-sin" term changes.

Several times before we have cheated a little for the sake of simplicity, and we have done it once again. To set things straight, let us note that the transformation matrix must be a square matrix with an order equal to the number of factors in the full factor matrix being rotated. If we wish to rotate clockwise the first and third factors in a five-factor matrix, the **T** matrix must have the form

$$\begin{bmatrix} \cos \theta & 0 & \sin \theta & 0 & 0 \\ 0 & 1 & 0 & 0 & 0 \\ -\sin \theta & 0 & \cos \theta & 0 & 0 \\ 0 & 0 & 0 & 1 & 0 \\ 0 & 0 & 0 & & 1 \end{bmatrix}$$

Also, although each rotation considers only a pair of factors, the effect of separate rotations of different pairs of factors is cumulative. Thus, after we are satisfied with the results of the rotations, we should be able to go directly from our initial factor loadings to our final factor loadings by using a single transformation matrix that contains these cumulative effects. This single transformation matrix provides a useful check on our successive computations and a compact way of transmitting our results to others; however, it cannot be computed without going through the successive pairwise rotations.

The final of cumulative transformation matrix is given by the successive product of the several pairwise transformations and has the general form

$$\mathbf{T}_c = \mathbf{T}_1 \mathbf{T}_2 \cdots \mathbf{T}_n \qquad (11.4)$$

For our data we had three rotations, and the $\mathbf{T}_c$ is

$$
\mathbf{T}_c = \begin{bmatrix} .9397 & .3420 & 0 \\ -.3420 & .9397 & 0 \\ 0 & 0 & 1 \end{bmatrix} \cdot \begin{bmatrix} .6428 & 0 & .7660 \\ 0 & 1 & 0 \\ -.7660 & 0 & .6428 \end{bmatrix} \cdot \begin{bmatrix} 1 & 0 & 0 \\ 0 & .8660 & -.5000 \\ 0 & .5000 & .8660 \end{bmatrix}
$$

$$
= \begin{bmatrix} .6040 & .6561 & .4523 \\ -.2198 & .6828 & .6967 \\ -.7660 & .3214 & .5567 \end{bmatrix}
$$

Multiplying the original factor matrix successively by the separate **T**s or by $\mathbf{T}_c$ yields the rotated factor matrix in Table 11-3. The pairwise plots of these factors appear in Figure 11-6a to c. From these plots and the table we see that all the variables lie in the same octant of the space, with the exception of 8, which is slightly outside. It might be possible through additional rotations to get all the variables to have all positive loadings.

**TABLE 11-3**

*Final Factor Matrix after 3 Graphic Rotations.*

| | Factor | | | |
|---|---|---|---|---|
| Variable | I | II | III | $h^2$ |
| 1 | .285 | .684 | .095 | .558 |
| 2 | .070 | .665 | .298 | .537 |
| 3 | .170 | .689 | .046 | .506 |
| 4 | .174 | .749 | .109 | .603 |
| 5 | .268 | .654 | .128 | .516 |
| 6 | .490 | .520 | .176 | .542 |
| 7 | .400 | .485 | .292 | .479 |
| 8 | .741 | .194 | −.107 | .600 |
| 9 | .446 | .320 | .395 | .457 |
| 10 | .657 | .239 | .012 | .489 |
| 11 | .640 | .086 | .379 | .561 |
| 12 | .574 | .151 | .534 | .616 |
| 13 | .180 | .112 | .702 | .538 |
| 14 | .069 | .145 | .685 | .496 |
| 15 | .364 | .198 | .586 | .515 |
| Eigenvalue | 2.706 | 3.179 | 2.126. | 8.013 |
| % of total | 18.0 | 21.2 | 14.2 | 53.4 |
| % of common | 33.8 | 39.7 | 26.5 | |

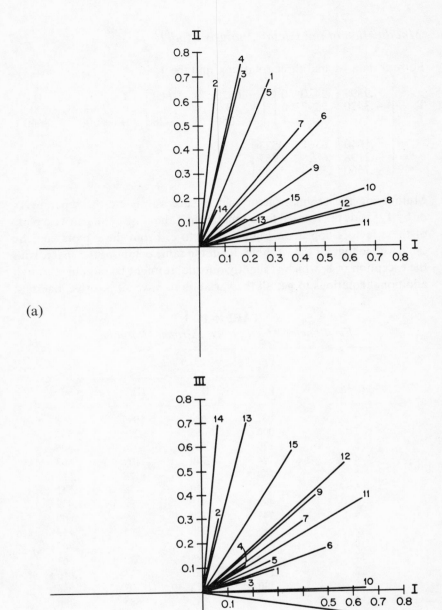

(a)

(b)

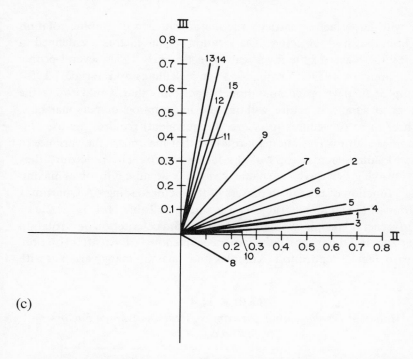

*Figure 11-6    Plots of factor pairs I–II (part a), I–III (part b), and II–III (part c) after final graphic rotation. Note that the three figures give the impression of a cone in the upper right octant of the 3-space.*

Such a finding would nicely reflect the fact that there were no negative correlations among the variables. For our present purposes it is sufficient to note that these results are not too dissimilar from those obtained by cluster analyzing these data in Chapter 9. Variables 1 to 5 make up factor II. Variables 8, 10, and 11 define factor I, and factor III is defined by 13 and 14. The other variables fall inside the cone. As is often the case, some of the variables are "factorially complex," that is, they have substantial loadings on more than one factor. If these factors are taken to indicate independent dimensions of ability, then the factorially complex tests require more than one ability. For example, variables 6 and 7 require the abilities indicated by factors I and II whereas variable 9 requires all three, and variables 11, 12, and 15 require abilities I and III.

   **Analytic rotation.**   Aside from the computational labor involved when computations must be done by hand, which may become considerable with large factor matrices, the major criticism of graphic rotation has been that it is subjective. The locations of the factors are defined in part by the biases of the investigator. In the early 1950s several people (e.g., Saunders, 1953; Carroll, 1953; and Neuhaus and Wrigley, 1954) attempted to develop computational algorithms that would rotate the factors to simple structure without the intervention of personal bias. Each of these procedures (they are all equivalent) requires the use of a computer and involves finding a rotation that maximizes the variance in factor loadings across rows of the factor matrix. The procedure they developed has been given the name *Quartimax* because it involves maximizing a function of the fourth powers of the factor loadings. A Quartimax rotation of the factors in Table 11-1 is shown in Table 11-4.

In general, people who used the Quartimax criterion for rotation found that it did not spread the variance out enough. Even after rotation, the early factors tended to be complex and have too many variables with

**TABLE 11-4**

*Matrix of Loadings after Quartimax Rotation of Three Factors from Table 11-1.*

| Variable | I | II | III | $h_2$ |
|---|---|---|---|---|
| 1 | .739 | .067 | .094 | .558 |
| 2 | .682 | .212 | −.160 | .537 |
| 3 | .712 | −.009 | −.003 | .506 |
| 4 | .775 | .044 | −.028 | .603 |
| 5 | .707 | .099 | .074 | .516 |
| 6 | .637 | .218 | .299 | .542 |
| 7 | .589 | .312 | .189 | .479 |
| 8 | .361 | .053 | .681 | .600 |
| 9 | .449 | .445 | .241 | .457 |
| 10 | .393 | .141 | .561 | .489 |
| 11 | .270 | .510 | .479 | .561 |
| 12 | .326 | .614 | .366 | .616 |
| 13 | .206 | .703 | −.047 | .538 |
| 14 | .209 | .655 | −.152 | .496 |
| 15 | .326 | .624 | .139 | .515 |
| Eigenvalue | 4.229 | 2.372 | 1.412 | 8.013 |
| % of total | 28.20 | 15.80 | 9.40 | |
| % of common | 52.78 | 29.60 | 17.62 | |

high loadings to provide a satisfactory result. The factors were difficult to interpret and did not conform very well to Thurstone's guidelines for simple structure.

A way to overcome the problems that plagued Quartimax while retaining its analytic (objective) character was proposed by Kaiser (1958), who reasoned that the variance in the matrix would be spread more evenly across the factors if the variance in loadings were maximized for the columns of the matrix rather than for rows. Such a criterion would force the rotation to maximize the number of very high and near-zero loadings and would minimize the number of loadings of moderate size. Kaiser gave his criterion the name Varimax, and, with a few modifications which need not concern us, this rotation algorithm has gained wide acceptance. In fact, the Varimax criterion is so generally available and so widely used that it has almost become a standard procedure to be applied to any factor analysis. Table 11-5 presents the results of a Varimax rotation of the three factors we have been considering.

**TABLE 11-5**

*Matrix of Loadings after Varimax Rotation of Three Factors from Table 11-1.*

| Variable | I | II | III | $h^2$ |
|---|---|---|---|---|
| 1 | .689 | .129 | .259 | .558 |
| 2 | .669 | .297 | .014 | .537 |
| 3 | .693 | .063 | .152 | .506 |
| 4 | .752 | .125 | .146 | .603 |
| 5 | .659 | .160 | .236 | .516 |
| 6 | .530 | .243 | .450 | .542 |
| 7 | .494 | .344 | .342 | .479 |
| 8 | .204 | .007 | .745 | .600 |
| 9 | .331 | .455 | .375 | .457 |
| 10 | .249 | .111 | .644 | .489 |
| 11 | .100 | .473 | .573 | .561 |
| 12 | .165 | .595 | .485 | .616 |
| 13 | .122 | .720 | .067 | .538 |
| 14 | .153 | .686 | −.040 | .496 |
| 15 | .210 | .633 | .266 | .515 |
| Eigenvalue | 3.260 | 2.514 | 2.238 | 8.013 |
| % of total | 21.73 | 16.76 | 14.92 | |
| % of common | 40.68 | 31.37 | 27.93 | |

A large number of other analytic rotation criteria have been proposed by various authors. Some are good and some are not so good, but none are as widely known and used as the two mentioned above. Some permit the factors to become correlated; others are, like Quartimax and Varimax, rigid orthogonal rotations. In general, the differences among them are not sufficiently large to be of interest to the "casual" user of factor analysis. Investigators who plan an extensive program of factor analytic research should become familiar with the various rotational procedures available by consulting texts on factor analysis (e.g., Harman, 1967) and the various journals (e.g., *Psychometrika* and *Multivariate Behavioral Research*) that publish theoretical articles on factor analysis.

**Comparison of factor solutions.** At this point we must confront a problem which has troubled factor analysts for many years, namely, the comparison of different factor solutions. This problem has several facets and numerous levels of complexity, of which our case is one of the least complex. We first consider the different facets, then the complexity dimension, and finally discuss our example.

The first issue to be faced is the grounds on which the comparison is to be made. Factor matrices may be compared on their scientific meaningfulness or usefulness. This is a comparison on the basis of the investigator's preference for one factor pattern rather than another, and, as such, can hardly be considered objective. This approach involves the investigator's ability to find a personally satisfying interpretation of the factors. He will tend to prefer rotations that fit his biases and which feel comfortable with his theories about the domain under study.

A second way in which solutions may be compared is on the basis of parsimony. One of the reasons why factor analysis has been widely used is because it reduces the number sources of variation. It can be argued that the best solution is the one which requires the fewest number of factors to account for some specified proportion of the variance in the set of variables. An alternative way to view this criterion is to say that for a given number of factors, the best solution is the one which accounts for the greatest proportion of the variance. If more factors are rotated than are to be interpreted, this criterion will always find the initial principal components solution to be superior. Otherwise, all rotations from a particular extraction will be equally good, and this criterion will be useful only for comparing extractions. (We discuss alternative extractions in the section on different initial solutions.)

The third criterion of comparison is the degree to which the solution approximates simple structure. This approach requires that we have some definition of simple structure that goes beyond Thurstone's (1947) verbal description. Various alternatives have been proposed, and none seems to be particularly satisfactory. One school of thought argues that a high number of "hyperplane loadings" (i.e., very small loadings) indicates the presence of a simple structure because then the variables will tend to have large loadings on relatively few factors. An alternative proposal (Thorndike, 1971) focuses on the number of high and low loadings. The index of simple structure of a single factor, $j$, is given by

$$g_j = 2\left[\sum_{i=1}^{n} \frac{(h_i - |a_{ij}|)^2 + a_{ij}^2}{\sum_{i=1}^{n} h_i^2}\right] - 1 \tag{11.5}$$

in which $h_i$ is the square root of the communality of variable $i$ and $a_{ij}$ is the loading of variable $i$ on factor $j$. The value

$$G = \frac{\sum_{j=1}^{m} g_j}{m}$$

($m$ = the number of factors) provides an index of simple structure for the whole matrix. This index seemed to have some utility in the study where it was proposed, but its properties have not been systematically investigated. Objective comparisons of factors on the basis simple structure have not progressed very far.

There is another way to compare factor solutions we cannot fruitfully apply to the three matrices above, but which holds promise for comparing different extractions. We noted earlier that each factor in a sense is a model for explaining the correlation matrix and that the more factors we use, the better the model will fit. Rotation of the factors does not change the reproduced correlations, but the different loadings from different initial extractions does. Thus it is possible to compare factoring procedures on the basis of the adequacy with which they reproduce the original correlations. The statistic involved, which may be used both as a test for how many factors to extract and as a means for comparing the solutions obtained by different extraction procedures, is discussed in the section on different initial solutions.

In addition to the different bases of comparison, there are three aspects of complexity that enter the factor comparison problem. The matrices being compared can differ in the type of rotation applied, the type of extraction procedure used, the sample from which the data were obtained, or any combination of these three. An added problem may arise when there are variables present in one analysis that are not present in the other. The difficulty is to determine how much of the difference is due to which source. It is particularly acute when trying to compare results published by different authors (e.g., the many published factorings of the California Psychological Inventory using different subjects, different extractions, and different rotations). The only real solution to this problem is for each author to publish his correlation matrix and a complete and accurate description of his factoring procedure so that others may reanalyze his data using their own extraction and rotation procedures. Then, if both used the same variables, differences in the factor matrices may be attributed to sample characteristics rather than to methodology. This type of complete reporting is distressingly infrequent, and its absence has almost certainly led to needless arguments and poor replications of studies.

After our extended digression into the difficulties of comparing factor matrices, we are ready to take another look at our three rotations. We avoid most of the issues of objective simple structure and focus instead on a logical analysis. (Those who see this as a copout are right. However, the objective indices of simple structure are not sufficiently well developed at this time to be appropriate for a discussion at the introductory level. The *G* index was mentioned above as an indication of the directions research in this area may take.)

The eigenvalues from the four solutions have been collected in Table 11-6. It can be seen from this table that each of the three rotations spreads

**TABLE 11-6**

*Eigenvalues of Three Factors from Four Solutions.*

| | Factor | | |
| Solution | I | II | III |
|---|---|---|---|
| Initial | 5.670 | 1.368 | .975 |
| Graphic | 2.706 | 3.179 | 2.126 |
| Quartimax | 4.229 | 2.372 | 1.412 |
| Varimax | 3.260 | 2.514 | 2.238 |

out the variance more than is the case for the initial solution. More importantly, however, note that Quartimax shows the tendency mentioned above to keep a large first factor. The graphic and varimax rotations do a much better job of distributing the variance across the three factors. This generally means that the factors will be less complex and easier to interpret.

Another way to compare these results is to look at the loadings of each factor individually. This has been done in Table 11-7 for the largest factor from each matrix (these are also the factors that have high loadings for the first six variables). Two things are readily apparent from this table. First, the graphic solution, which was obtained from an inspection of the factor plots and was guided by a human logic approach to simple structure, is much more similar to the Varimax factor than it is to the Quartimax. This means that Varimax came much closer to the author's logical rotation than did Quartimax and may be considered superior in that regard. Second, there are more loadings near zero for the Graphic and Varimax rotations than there are for Quartimax, a feature which may also be taken to indicate better simple structure. A comparison of the other factors

**TABLE 11-7**

*Loadings of 15 Variables for the Largest Factor from Each of Four Solutions.*

| Variable | Initial (I) | Graphic (II) | Quartimax (I) | Varimax (I) |
|---|---|---|---|---|
| 1 | .664 | .684 | .739 | .689 |
| 2 | .614 | .665 | .682 | .669 |
| 3 | .576 | .689 | .712 | .693 |
| 4 | .646 | .749 | .775 | .752 |
| 5 | .649 | .654 | .707 | .659 |
| 6 | .717 | .520 | .637 | .530 |
| 7 | .691 | .485 | .589 | .494 |
| 8 | .526 | .194 | .361 | .204 |
| 9 | .658 | .320 | .449 | .331 |
| 10 | .559 | .239 | .393 | .249 |
| 11 | .615 | .086 | .270 | .100 |
| 12 | .678 | .151 | .326 | .165 |
| 13 | .500 | .112 | .206 | .122 |
| 14 | .447 | .145 | .209 | .153 |
| 15 | .615 | .198 | .326 | .210 |

from Tables 11-3, 11-4, and 11-5 will reveal a like similarity and superiority for the Varimax and Graphic solutions.

## COMMUNALITY AND NUMBER OF
## FACTORS PROBLEMS REVISITED

We have seen that the investigator is usually interested in rotating an obtained factor matrix to clarify the interpretation of the factors, but at this point we have not disentangled the problem of the number of factors retained from the problem of what to use for communality estimates. Before proceeding to some special factoring methods that to some extent separate these problems, it may be well to consider the twin problems from the perspective of the matrix to be interpreted.

The logic of the Kaiser criterion is very appealing. It makes good sense to retain only those factors that account for at least as much variance as a simple variable. However, because the rotated factors are to be interpreted, it is also appealing to apply the Kaiser criterion to the rotated matrix. We might use as a starting point for rotation the principal components matrix and rotate one more factor than is called for under the Kaiser criterion (call this number $K$). If all $K$ rotated factors have eigenvalues greater than 1.0, we rotate $K + 1$ factors and check the result. Should these $K + 1$ rotated factors still have eigenvalues greater than 1.0, we may try $K + 2$, $K + 3$, and so on until one rotated factor eigenvalue (quite possibly not the last one) drops below unity. We may then back up to the previous solution and accept that result (say, $K + p$ factors) as our number of factors. In general, this decision rule will retain several more factors than the original Kaiser criterion.

Two things happen when the Kaiser criterion is applied to the rotated factor matrix rather than to the original factor matrix. First, we have the more satisfying situation of applying the criterion to the factors to be interpreted, and second, we run less risk of retaining too few factors. Several investigators (e.g., Cliff and Hamburger, 1967; Levonian and Comrey, 1966) have warned that serious distortions of the factors may result if not enough factors are rotated. The clarity of the factor structure can be seriously impaired if there is not enough space for the variance to spread out. The result is likely to be a few large complex factors that defy interpretation as meaningful hypothetical constructs.

In the absence of definite expectations about the number of factors to expect in a given set of data (and this is usually the case), it is often useful to experiment with various procedures in arriving at a final factor matrix. Although such muddling around with the data often horrifies statistical purists, such practices can lead the investigator to a better understanding of his domain of variables. We have just discussed one way of muddling with the number of factors. Now, let us use our $K + p$ factors as a logical solution to that problem and proceed to iterate for the communalities. That is, take the observed communalities from $K + p$ factors and insert them in the main diagonal of **R**. We may now factor our reduced matrix, obtain new estimates of the communalities for $K + p$ factors, and repeat the process until the communalities stabilize.

At this point the question of how many factors to retain for rotation and/or interpretation can again be asked. In one sense we already have $K + p$ as the answer. However, the fact that this decision rule was applied to an unreduced correlation matrix, whereas we are now dealing with a matrix that has been reduced by some (perhaps substantial) amount, may cause the investigator to wish to reconsider. Alternatively, the investigator may have started with reliabilities or squared multiple correlations as communality estimates and may wish a decision rule without regard to iteration. In either case, we can modify the logic of the Kaiser criterion for the new situation. At this point the communalities can be considered determined. Therefore, the *average communality* is the common variance of the average variable. Thus we can argue that those rotated factors whose eigenvalues exceed the average communality for a variable are the ones that warrant interpretation.

The steps outlined above involve several rotations, several refactorings in the iteration process, and several more rotations. Obviously, such a sequence of analyses will require substantial amounts of computer time and, depending on the scope and purpose of the particular study, may not be worth the effort and expense. We might note, however, that the three steps are largely independent. The investigator who desires a rotated principal components solution may wish to retain $K + p$ factors. Iteration for communalities is possible with any number of factors up to $n - 1$, and the investigator who has communality estimates may wish to retain only those factors which account for a greater proportion of the common variance than does the average variable.

## DIFFERENT INITIAL SOLUTIONS

Up to this point our discussion of factor analysis has focused on the principal-axis factoring of an unreduced correlation matrix (one with ones in the main diagonal). Mention was made of the use of various other values in the main diagonal as communality estimates, but the logic of the analysis remained the description of the relationships observed in the given matrix of correlations. Various authors have extended the logic of factor analysis to include other objectives. These extensions are of two kinds, statistical objectives and psychometric objectives. The factoring procedures described in the remainder of this section by no means exhaust the range of possibilities that have been proposed, but they can be taken as examples of the kinds of proposed alternatives. In a sense, they are not really alternative methods of factor analysis but rather different things which can be done, under certain sets of assumptions, to the correlation matrices before factoring to achieve different purposes.

The distinction between statistical and psychometric factoring procedures was proposed by Kaiser and Caffrey (1965) and involves the nature of the assumptions one is willing to make about his data. Statistical factoring involves an assumption that the variables in the study constitute a population and the subjects on whom measurements were taken constitute a sample. The task is to find the factor matrix in the sample data that is most like that which exists in the population. This is a problem in statistical estimation very similar to that encountered in univariate statistics when, for example, one wishes to get the best possible estimate of a population mean and standard deviation. These best estimates of population values are called *maximum likelihood estimators*, and factoring procedures that produce maximum likelihood estimates of the factor loadings are classed as statistical factorings.

Psychometric factor analysis, on the other hand, requires the assumption that the subjects in the study are a population and that the variables have been sampled from a universe of possible variables. The problem is to estimate the factors that exist in the population from the data on the sample or to determine the factors that have the greatest generality in the population of variables.

**Statistical factoring.** The application of maximum likelihood logic to factor analysis is due to Lawley (1940), who first derived the necessary equations. Unfortunately, these equations could not be solved by direct computation; thus it was necessary to develop an iterative and approximate solution. Several authors have proposed different iterative solutions to Lawley's likelihood equations, notably Lawley himself (1943) and Rao (1955), whose procedure was called canonical factor analysis because it maximizes the canonical correlation between the set of factors and the set of variables. However, the best presently available solution has been provided by Jöreskog (1967). This procedure involves the iterative solution of a set of complex matrix equations based on the assumption of a particular number of factors. Given the number of factors, the correlation matrix is used to estimate the uniquenesses of the variables (which is also an estimate of their communalities). Factors are then extracted from the reduced correlation matrix.

As we have seen, the communalities of the variables depend on the number of factors extracted. When the reduced correlation matrix is factored, the loading of a particular variable on a particular factor will vary, depending on the communality estimates that have been placed in the main diagonal. Thus it is important to know the correct number of factors to use because, unlike principal components analysis, the loading of a variable on the first factor (and all other factors) will change, depending on the number of factors that follow it. Fortunately, a solution to this problem is available in the form of a test of statistical significance of the factors. The test involves the adequacy with which the maximum likelihood solution for the particular number of factors reproduces the original correlation matrix. It tests whether the model represented by this number of factors is adequate to describe the sample data.

The test to be used *with a maximum likelihood analysis* is given by the equation

$$\lambda = (N-1)\log_e |\mathbf{R^*}| - \log_e |\mathbf{R}| + \sum_{i=1}^{n}\sum_{j=1}^{n} r^{*ij} r_{ij} - n \qquad (11.6)$$

where $|\mathbf{R^*}|$ is the determinant of the reproduced matrix,

$|\mathbf{R}|$ is the determinant of the original correlation matrix,

$r^{*ij}$ are the elements of the inverse of $\mathbf{R^*}$,

$r_{ij}$ are the elements of $\mathbf{R}$, and

$n$ is the number of variables.

The statistic $\lambda$ has an approximate $\chi^2$ distribution with $1/2[(n - m)^2 - n - m]$ degrees of freedom. If the probability of the $\chi^2$ variable is less than .01 (i.e., it is statistically significant) then the fit of the model is poor, and more factors should be used. A nonsignificant value of the statistic means that the fit is acceptable and enough (or perhaps too many) factors have been used. The best way to use the test is to start with a small number of factors and add one factor on successive refactorings until the statistic becomes nonsignificant. The last factor to be added before the statistic turns nonsignificant is the last significant factor. Used in this way, equation (11.6) is a test of the statistical significance of each factor in turn.

An example of the kind of results which may be obtained from Jöreskog's (1967) maximum likelihood procedure is presented in Table 11-8 and the Varimax rotation of these factors is in Table 11-9. For these data with a sample of 310 the test statistic became nonsignificant after inclusion of the fifth factor.

One feature is quickly apparent from these tables. Even five maximum likelihood factors do not account for as much of the variance in the original variables as do the first three principal components. This loss of variance is a universal feature of analyses of reduced correlation matrices and is, in a sense misleading. The use of common variance rather than total variance in the main diagonal of the correlation matrix means that there is less total variance available for analysis. However, we may say with certainty that the average factor loading will be larger for a principal component analysis than for any other factoring of a given set of data.

Another feature of Table 11-8 requires our attention. The communialities for variables 7 and 9 are both 1.00, and both variables have identical factor loadings. There are a few occasions (of which this is one) when the factoring of a reduced correlation matrix yields communialities of 1.00 or greater. Such occurrences are known as Heywood cases and appear when the reduction of the correlation matrix makes the determination of proper communialities impossible. Of course, a Heywood case is, in a sense, a meaningless result, because it is not possible for a set of factors to account for more than all a variable's variance.

There are various ways of handling Heywood cases when they occur. One is simply to use fewer factors. However, as in the present case, this solution may involve discarding significant factors. Jöreskog's (1967) solution for maximum likelihood factors is to analyze the matrix of partial variances and covariances with the offending variable(s) removed. Then one factor is used for each removed variable and the loadings of the vari-

## TABLE 11-8
*Unrotated Maximum Likelihood Factor Matrix for 15 Variables from Table 9-1.*

| Variable | 1 | 2 | 3 | 4 | 5 | $h^2$ |
|---|---|---|---|---|---|---|
| 1 | −.462 | −.067 | −.436 | .242 | −.002 | .466 |
| 2 | −.492 | −.038 | −.301 | .172 | −.188 | .398 |
| 3 | −.316 | −.058 | −.496 | .267 | −.028 | .421 |
| 4 | −.462 | −.029 | −.435 | .321 | −.093 | .516 |
| 5 | −.392 | −.106 | −.514 | .191 | −.035 | .466 |
| 6 | −.527 | −.096 | −.434 | .052 | .093 | .487 |
| 7 | −.854 | −.520 | 0.000 | 0.000 | 0.000 | 1.000 |
| 8 | −.351 | −.038 | −.359 | −.031 | .397 | .412 |
| 9 | −.854 | .520 | 0.000 | 0.000 | 0.000 | 1.000 |
| 10 | −.410 | .096 | −.319 | −.013 | .179 | .311 |
| 11 | −.462 | −.010 | −.339 | −.300 | .132 | .437 |
| 12 | −.486 | .029 | −.432 | −.371 | .019 | .561 |
| 13 | −.328 | .000 | −.331 | −.340 | −.244 | .392 |
| 14 | −.339 | .077 | −.220 | −.206 | −.157 | .444 |
| 15 | −.421 | .019 | −.405 | −.293 | −.122 | .444 |
| Variance | 3.801 | .589 | 2.028 | .774 | .361 | 7.548 |

## TABLE 11-9
*Rotated Factors for Maximum Likelihood Solution in Table 11-8.*

| Variable | 1 | 2 | 3 | 4 | 5 |
|---|---|---|---|---|---|
| 1 | .120 | .141 | .598 | .166 | .218 |
| 2 | .194 | .195 | .512 | .245 | .025 |
| 3 | .021 | .023 | .608 | .137 | .177 |
| 4 | .158 | .114 | .664 | .147 | .123 |
| 5 | .017 | .097 | .602 | .225 | .209 |
| 6 | .114 | .205 | .478 | .283 | .351 |
| 7 | .152 | .870 | .339 | .253 | .202 |
| 8 | .062 | .091 | .246 | .132 | .567 |
| 9 | .887 | .146 | .229 | .313 | .201 |
| 10 | .209 | .048 | .284 | .217 | .371 |
| 11 | .121 | .153 | .148 | .465 | .400 |
| 12 | .122 | .105 | .190 | .614 | .351 |
| 13 | .044 | .072 | .152 | .600 | .039 |
| 14 | .160 | .066 | .145 | .427 | .058 |
| 15 | .096 | .083 | .233 | .581 | .187 |
| Variance | 1.015 | .963 | 2.486 | 1.959 | 1.125 |

ables on that factor are computed. In our example the two offending variables are placed wholly within the plane defined by the first two factors such that the first factor falls exactly between them and they define a bipolar second factor. The loadings for the other variables for the first two factors are their correlations with these factors, as defined by the Heywood case variables. The third, fourth, and fifth factors are independent of the first two (as always), and because the two offending variables fall completely within the plane of factors I and II, they have zero loadings on the last three factors. The loadings of the other variables on the last three factors are found from the matrix of second-order partial correlations, which is the residual correlation matrix after the first two factors have been removed. The initial factor matrix is somewhat improved by the Varimax rotation, in that the general first factor is replaced by a factor more precisely located at variable 9 and the second factor takes care of variable 7. The last three factors show a striking resemblance to the Varimax factors in Table 11-5.

Another statistical factoring procedure has been proposed by Harman (1967). Although its logical starting point and computational procedures are different from those of Lawley and Jöreskog, the results are quite comparable. Harman started with the goal of finding those factors that would give the best reproduction of the original correlation matrix with communalities in the main diagonal. He termed the method minimum residual factor analysis (*minres*), because it involves solving iteratively for a set of communalities *for a given number of factors* that yield the smallest possible residual correlations.

Because the minres communalities depend on the number of factors, it is obvious that a test of statistical significance of the factors such as that described above would be highly desirable. Harman has advocated the use of Rippe's (1953) modification of the Lawley maximum likelihood test. The only change is in the degrees of freedom, which become $1/2[(n - m)^2 + n - m]$ in the Rippe modification. Otherwise, the two tests are identical and can be used in the same way.

There is an interesting little section in Harman's treatment of minimum residual factor analysis (1967, p. 195) which nicely illustrates the close similarity between minres and maximum likelihood factoring. He shows that a principal axis factoring of the correlation matrix with minimum residual communalities (i.e., communalities determined to optimize his criterion) yields a solution that is equivalent to the minres solution.

Also, a principal axis factoring of the original correlation matrix with maximum likelihood communalities is equivalent to the maximum likelihood solution. That is, both procedures involve nothing more than different ways to estimate the communality. In practice, the results from the two procedures are, in general, quite comparable for a given number of factors.

One important difference between minres and maximum likelihood is the difference in the degrees of freedom associated with their statistical tests for the number of factors. The consequences of this difference are illustrated by a comparison of the minres factoring of our example data in Table 11-10 with the maximum likelihood results. The greater degrees of freedom associated with the Rippe test causes us to use three factors rather than five. (Note that we do not encounter a Heywood case here, although Harman and Fukuda, 1966, have provided an iterative solution for the problem.) MacDonald (1975) has questioned the appro-

**TABLE 11-10**

*Unrotated and Rotated Minimum Residual Factor Matrices for*
*15 Variables from Table 9-1.*

| Variable | Unrotated | | | | | Rotated | | |
|---|---|---|---|---|---|---|---|---|
| | 1 | 2 | 3 | $h^2$ | | 1 | 2 | 3 |
| 1 | .636 | −.272 | .014 | .479 | 1 | .621 | .189 | .239 |
| 2 | .580 | −.177 | .220 | .417 | 2 | .569 | .302 | .041 |
| 3 | .541 | −.288 | .022 | .376 | 3 | .570 | .124 | .188 |
| 4 | .622 | −.333 | .095 | .507 | 4 | .675 | .168 | .154 |
| 5 | .616 | −.230 | .014 | .432 | 5 | .578 | .207 | .235 |
| 6 | .689 | −.092 | −.099 | .493 | 6 | .504 | .305 | .382 |
| 7 | .656 | −.027 | .015 | .431 | 7 | .465 | .375 | .273 |
| 8 | .496 | −.010 | −.390 | .398 | 8 | .244 | .138 | .565 |
| 9 | .620 | .114 | .033 | .398 | 9 | .346 | .461 | .257 |
| 10 | .516 | .013 | −.173 | .297 | 10 | .296 | .249 | .383 |
| 11 | .585 | .285 | −.148 | .445 | 11 | .159 | .492 | .421 |
| 12 | .659 | .351 | −.037 | .559 | 12 | .191 | .625 | .363 |
| 13 | .467 | .309 | .203 | .354 | 13 | .150 | .573 | .059 |
| 14 | .408 | .226 | .161 | .244 | 14 | .158 | .464 | .062 |
| 15 | .582 | .267 | .095 | .419 | 15 | .230 | .571 | .201 |
| Variance | 5.102 | .797 | .351 | 6.250 | | 2.724 | 2.244 | 1.282 |

priateness of the Rippe test, and in light of the similarity between the two factoring procedures, it may be appropriate to use Lawley's degrees of freedom with minres factoring. Allowing for the difference in the number of factors, the three minres factors are quite similar to the third, fourth, and fifth maximum likelihood factors.

**Psychometric factoring.** The ideas behind psychometric factoring are relatively new, being due mainly to the work of Kaiser. In advocating what has become known as the Kaiser criterion he showed (Kaiser, 1960) that to guarantee positive generalizability between the factors found from a principal components factoring of a sample of variables and the population of variables, it is sufficient that the factors have eigenvalues greater than 1.0. Taking off from this point, Kaiser and Caffrey (1965) developed *alpha* factor analysis. They showed that by rescaling the correlation matrix in terms of the common portions of the variables using the equation

$$\mathbf{A} = \mathbf{H}^{-1}(\mathbf{R} - \mathbf{U}^2)\mathbf{H}^{-1}$$

where **H** is a diagonal matrix of "common part standard deviations" (p. 8) it is possible to obtain factors that have a positive coefficient $\alpha$ (in the Cronbach, 1951, sense) with the universe common factors. Application of this rule always results in the same number of factors as would be obtained by applying the Kaiser criterion to a principal components analysis of **R**, but the eigenvalues will be smaller.

The results of an alpha factoring of our data are presented in Table 11-11. We see that the two factors account for a relatively small amount of the total variance and that the rotated factors are not as clearly defined as has been the case when three factors are used. In general, these factors do not appear to be very satisfactory.

**Choice of method.** When considering the use of an alternative method of factor analysis one must evaluate the assumptions involved. A principal components analysis is based on the assumption that both variables and subjects constitute populations. Thus for the data at hand the solution is exact and provides a least-squares fit. This is desirable, but it severely limits our ability to claim that the results can be generalized to new groups of subjects or different variables. The statistical factoring procedures, with their assumption that the variables are fixed and subjects are sampled, permit us to generalize our results *for these variables* to a new sample of subjects. The assumption that subjects have been sampled is fairly easy to accept and fairly useful. On the other hand, psychometric factoring

**TABLE 11-11**
*Unrotated and Rotated Alpha Factor Matrices for 15 Variables
from Table 9-1.*

| Variable | Unrotated | | | Rotated | |
|---|---|---|---|---|---|
| | 1 | 2 | $h^2$ | 1 | 2 |
| 1 | .63 | −.31 | .49 | .67 | .21 |
| 2 | .60 | −.18 | .40 | .56 | .29 |
| 3 | .57 | −.32 | .43 | .64 | .16 |
| 4 | .65 | −.37 | .56 | .73 | .18 |
| 5 | .65 | −.26 | .49 | .65 | .26 |
| 6 | .70 | −.15 | .51 | .61 | .37 |
| 7 | .66 | −.06 | .44 | .52 | .41 |
| 8 | .48 | −.07 | .24 | .40 | .28 |
| 9 | .62 | .09 | .39 | .39 | .49 |
| 10 | .52 | −.06 | .27 | .42 | .32 |
| 11 | .56 | .21 | .35 | .26 | .53 |
| 12 | .63 | .28 | .47 | .26 | .64 |
| 13 | .43 | .28 | .26 | .12 | .50 |
| 14 | .39 | .21 | .20 | .13 | .42 |
| 15 | .55 | .25 | .37 | .23 | .56 |
| Variance | 5.10 | .78 | 5.88 | 3.48 | 2.41 |

requires that we assume our subjects are a population. The factors we obtain from our sample of variables will be maximally generalizable to a new sample of variables *for the same subjects.* This type of generalization is seldom of interest. Therefore, it would seem that psychometric factor analysis can only be used appropriately in a few, unusual situations.

## SUMMARY

This chapter has covered various modifications of the general factor model. Modifications generally take the form of analyzing some reduced correlation matrix or of transforming the location of the initial factors by a rotation procedure. We have discussed several criteria for the adequacy of factor solutions, including ability of the factors to reproduce the original correlations, minimum rank of the correlation matrix, and determination of the "correct" or optimum number of factors by using statistical criteria such as those of Bartlett and Lawley and psychometric criteria proposed by Kaiser and Cattell. The problem of communality estimates was then discussed, with attention being given to direct iteration, the use of squared multiple correlations, and reliabilities.

The choice of alternative locations for the factors, known as factor rotation, was then considered. After a discussion of the reasons for rotation and Thurstone's criteria for acceptable solutions, including the logic of simple structure, the principles of rotation were discribed with graphic examples. Then two analytic rotation criteria, Quartimax and Varimax, were illustrated and compared with the graphic solution. Finally, three alternative initial solutions were presented. Maximum likelihood and minimum residual procedures basically involve the determination of communalities in such a way as to maximize our ability to generalize the factors obtained to new samples of subjects and to reproduce the original correlations, respectively. The two methods are usually quite similar in their results. Alpha factor analysis determines factors with maximum generalizability to a new sample of variables. For most applications one of the statistical factoring methods would seem preferable.

# 12

## OTHER ISSUES IN
## FACTOR ANALYSIS

The preceding chapters in Parts II and III of this book have dealt with the logic and, to a small extent, with the mathematics of selected internal and external factor analysis procedures. Our coverage of these topics has focused on the most commonly used methods and has ignored many useful types of multivariate analysis. For example, we have not discussed multivariate analysis of variance (MANOVA). Nor have we considered the various methods of multidimensional scaling that have recently been developed. MANOVA, which tests the differences between group means for sets of variables considered simultaneously, is discussed by Cooley and Lohnes (1971). The reader interested in scaling individuals or objects on several traits should consult the work of Shephard (1974) and Carroll and Chang (1970).

In this chapter we will clean up some loose ends from the techniques we have discussed and go on to consider experimental procedures and cautions in the use of these procedures. We begin with a brief look at an exciting new development in the area of internal factor analysis, a technique called confirmatory factor analysis. We then go on to consider the determination of scores for individuals on internal factors.

Next, the interpretation of factors is discussed, and we conclude this chapter with a consideration of some of the pitfalls that await the investigator who sets out to perform a multivariate study.

## CROSS-VALIDATION IN FACTOR ANALYSIS

One of the criticisms frequently leveled at internal factor analysis is that there is no way to determine how well the factors found in one set of data will fit the data from a new set of observations. As we have seen, this is not a problem for external factor analysis. The generalizability of a set of regression weights can be tested by cross-validation. If the relationships between the composites in the two sets of variables, be they from multiple regression, canonical analysis, or discriminant function, remain satisfactorily high in a new sample, then we may conclude that the composites or factors we have found are generalizable.

The problem is not so simple in internal factor analysis, and the difficulty arises because there is no external variable or set of variables to provide an objectively defined "best position" for the factors. Nor is there an objective index of the degree of fit which the cross-validation correlation provides for external factor analysis. The problem is compounded by the fact that internal factors can be rotated without affecting the mathematical quality of the solution.

Several attempts have been made to overcome this problem so that the results from different sets of data may be compared. One approach has been to develop indices of factor similarity such as the root-mean-square and coefficient-of-congruence indices described by Harman (1967, pp. 268–272). These procedures by themselves are of little real value because they are affected by (and dependent on) the rotation solution that has been chosen.

A more fruitful line of attack has been taken by Ahmavaara (1954) and further pursued by Cattell and his co-workers (Cattell and Baggaley, 1960; Cattell and Muerle, 1960; Eber, 1966, 1968) and others. This approach involves rotation of the factors, but it puts constraints on the rotation such that the resulting factors will be most similar to some desired solution. Ahmavaara's procedure rotates simultaneously the factors from two sets of data to positions of maximum congruence. This may maximize the similarity between factors, but it does not take simple structure or interpretability into account. Therefore, the matched factors may not be satisfactory.

An alternative rotational approach advocated by Cattell has been given the name "Procrustes rotation" by him. It actually includes a variety of methods for rotating an initial factor matrix to maximum agreement with some predefined criterion structure. This is not specifically a method for judging the generalizability of the factors. However, if two different sets of data are collected and both are rotated to maximum fit with the same criterion set of factors, the result is, in effect, a cross-validation of each of the sets of factors. If it is possible to obtain a high degree of similarity between the factors from the two independent sets of data, the factors can be judged to have generality. Of course, an alternative way to use the Procrustes approach, one which is more directly cross-validational in nature, is to use the factors from one sample as the criterion matrix and to rotate the factors from the second sample to positions of maximum similarity with those of the first.

What is probably the best currently available method for performing something akin to cross-validation for internal factor analyses is a procedure developed by Jöreskog (1969) and called by him confirmatory factor analysis. Once again, a criterion or hypothesis factor matrix is postulated, and the data are analyzed to provide the best possible approximation to the preset criterion. However, in this approach the factors are *extracted* under these constraints rather than being rotated to agreement. The procedure is a maximum likelihood one and has the advantage of providing a test of statistical significance of the goodness of fit of the data to the hypothesis matrix. In addition, it is possible to vary the number of constraints on the data. All or only some of the factor loadings can be specified, and the correlations among the factors can also be set, making it possible to treat oblique as well as orthogonal solutions. When some of the factor loadings are left free to vary, maximum likelihood estimates of the unspecified loadings are obtained. This development by Jöreskog has opened some exciting possibilities for programmatic research in factor analysis, which are discussed shortly.

## FACTOR SCORES

At several places in our discussion of external factor analysis and of cluster analysis we have seen that the calculation of scores for individuals on the composites presents no particular problem. All that is necessary is to multiply each individual's standard score on each variable by the standard partial regression weight ($\beta$ weight) or its appropriate equivalent and then sum these across variables. It is possible to do essentially the same thing in the case of classical internal factor analysis in its various forms; however, the task is a little more complex.

The cause of the difficulty was foreshadowed in our discussion of factor analysis as multiple regression in Chapter 10. In the case of external factor analysis, composites of the variables in one set are used to predict composites of the variables in the other. There are observed variables on both sides, and the composite scores can be computed directly from them. In the development of internal factor analysis, however, we have a set of observed variables being predicted by a set of latent variables, the factors. Thus we have observed variables on one side and unobserved variables, which are doing the predicting, on the other. It is, therefore, necessary to estimate the scores on the factors subject to the restriction that they do the best possible job of predicting the observed variable scores. This may be accomplished for the case of orthogonal factors by using the equation (see Harman, 1967, p. 351)

$$\hat{\mathbf{F}} = \mathbf{S}'\mathbf{R}^{-1}\mathbf{Z}' \tag{12.1}$$

where $\hat{\mathbf{F}}$ is the $m \times N$ matrix of estimated common factor scores of $N$ people on $m$ factors, $\mathbf{S}'$ is the transpose of the matrix of orthogonal factor loadings, $\mathbf{R}^{-1}$ is the inverse of the correlation matrix of the observed variables, and $\mathbf{Z}$ is the matrix of standard scores of the $N$ people on the $n$ observed variables. Looking at it in another way, the product $(\mathbf{S}'\mathbf{R}^{-1})$ yields the necessary matrix of $\beta$ weights.

Factor scores are useful in two situations which derive from a basic *raison d'etre* of factor analysis, the reduction of the number of sources needed to explain observed variation. First, factor scores are used to reduce the redundancy in a set of observed scores. For example, there are many psychological measuring instruments which report scores on a variety of scales (e.g., the California Psychological Inventory, Gough, 1956). When correlations among the scales are relatively high, scores on the separate scales are not independent and represent substantial repetition of information. If a factor analysis of these scales shows that most of their variance can be accounted for by a relatively small number of factors, interpretation of the instrument may be enhanced by using this smaller set of independent dimensions.

The second use to which factor scores are put is to reduce the number of variables that must be considered at later stages of an investigation. For example, an investigator may collect data on a large number of variables, such as was the case in the Project TALENT research (Flanagan, 1962). Cooley and Lohnes (1971) describe how sets of these variables were factor analyzed, factor scores computed, and then group comparisons (e.g. between males and females) made on the factor scores. The number of variables in this study is so great that even with the large samples used, consideration of individual variables did not appear promising. By using a relatively small number of factor scores it was possible to conserve degrees of freedom and perform analyses that would not have been possible otherwise.

At this point it seems advisable to bring up a semantic distinction which, although not particularly crucial, will serve to alert the reader to erroneous statements that appear with distressing frequency in the research literature. The term "factor score" should be used to refer to estimates of the individuals' scores on the factors as they are developed by equations such as (12.1). All too often one finds on close examination of an author's data that the variables have been factor analyzed and each variable has been assigned, *in toto*, to the factor on which it has its highest loading (or, still worse, we occasionally find a high loading overlooked and a variable assigned to another factor because the author is more concerned with a pet interpretation than with what his data show). Raw scores or, perhaps, standard scores are then summed for the variables assigned to the factor, and the result is called a factor score. Although in the first instance this is not particularly harmful so long as standard scores are used (assuming that a correlation matrix has been analyzed), in the second case the results are downright misleading. In either case, the resulting scores are not factor scores. They are essentially cluster scores of the type described in Chapter 9 and have the property that the correlations among them are unknown. This may not present a serious problem for the first use of factor scores described above, but it will distort the results of subsequent statistical procedures applied to the scores, because the sets of scores will not be statistically independent or correlated to a known degree. In any case, the clarity of communication will be enhanced if the term factor score is reserved for its appropriate use and if we adopt some other term, such as cluster score or composite score, where the scores are not strictly factor scores.

## APPLICATION OF FACTOR ANALYSIS

With the possible exception of multiple correlation, factor analysis, as it has been discussed in the preceeding two chapters, is by far the most commonly used way to analyze multivariate data in psychology and education. The research literature contains several thousand studies that have used factor analysis in one way or another (see Bolton, 1973, for an annotated bibliography). However, methodologically proper applications are more difficult to find. This occurs largely because there are so many choices open to the investigator after he has collected his data. Tests of mean differences that are performed by analysis of variance procedures require that the study be planned carefully in advance so that questions of interest will be answered. Otherwise, certain tests cannot be performed, and it is easy to tell that insufficient care and planning went into the investigation. Factor analysis contains no such safeguards, and the result has been a large number of poorly conducted studies yielding results whose interpretation is highly questionable.

Leading proponents of factor analysis have advocated and, in general, practiced, good factor analytic methodology for many years. The work of Cattell and Guilford over the years can be taken as examples solid experimental principles being applied in a consistent, well-planned program of research. Scientists such as these, although they disagree with each other on many issues and are themselves centers of controversy, nevertheless exemplify in their work some of the best uses of factor analysis as a tool of science.

The real problem has arisen as a direct result of readily available computer programs for factor analysis. This has led to what we might call the "Saturday-night-special" factor analysis, in which the investigator with little or no understanding of factor analysis takes a deck of data cards over to his computer consultant and asks for a factor analysis. Often, the consultant is left to make decisions about communalities, number of factors, rotation, and other program options, with little understanding himself of the consequences of his choices and no knowledge of the nature of the study that produced the data. (Or, worse yet, the data may be a conglomeration of variables from various sources.) It is little wonder, then, that many factor analyses add little substantive information about the field of study. Factor analysis requires the same careful attention to the details of experimental method that is necessary for any other study.

**Planning the study.** The place where the first decisions about factoring procedures should be made is in the early stages of planning. The investigator should decide that factor analysis will or will not be used before he selects the variables for his study. Often, as is the case when a particular instrument is being investigated, this cannot strictly be done, but in these cases, other instruments will also be included. These latter variables should then be selected with a view to their ability to clarify the nature of the instrument of primary interest. For example, if one has developed a new measure of certain dimensions of personality that are hypothesized to be largely independent of one another, it would be well to include in the study established personality measures with known properties so that the relationships between the new instrument and others can be evaluated.

Variables with known properties are called *marker variables* and serve two purposes. First, they aid in the interpretation of the factors. The investigator knows (supposedly) what his instrument is designed to measure. If two or three other variables that are considered to be measures of the same dimensions are included for each dimension measured by the new instrument, then finding that the scales of the new instrument have high loadings on the same factors as the appropriate marker variables is fairly strong evidence of the validity of the new scales as measures of those dimensions.

A second advantage of including variables with known factorial content in a study of a new instrument is that the marker variables tie the new variables in to previous research. The literature is full of one-shot studies of instruments. Such studies add little to our understanding, because there is no overlap among them. The best approach is an interlocking series of studies in which a few variables are altered or added from one investigation to the next. However, single studies can be placed in an overall research context by the inclusion of marker variables.

Another feature of experimental design which is often overlooked is the selection of subject populations. What we characteristically find in the literature is what may be called samples of convenience in which there has been no particular attention paid to important dimensions of variation in subjects. The worst example of this error is the investigator who grabs a bunch of data lying around and rushes off to analyze it. Important sample characteristics such as age, sex, and socioeconomic status should be systematically varied in a series of studies or, at least, held constant and be known. Once again, planning and careful selection are necessary.

**Conducting the study.**    There are three issues which warrant discussion at this point. The first is a decision regarding sample size. In general, we can say that the larger the sample the better, keeping in mind the point made above. In inferential statistics excessively large samples lead to conclusions that very small differences are statistically significant; thus some authorities (e.g., Hays, 1963; Kendall, 1959) recommend using samples of moderate size. This is seldom a problem in factor analysis. Indeed, the problem is more likely to be that we obtain large factors containing error covariation than that important relationships will go undetected or small relationships will be found "significant." The amount of variance available for analysis in a correlation matrix of 20 variables is the same regardless of whether the correlations are based on a sample of 50 or 5000. However, large samples are more likely to give us accurate estimates of population values and will therefore generally yield better, more stable results.

Of course, large samples can be taken to extreme degrees. When statistical tests such as the Bartlett test or those advocated by Jöreskog (1967, 1969) and Rippe (1953) are not employed, large sample size will not be a problem, because tests for the number of factors such as the Kaiser criterion do not take $N$ into consideration. However, if statistical significance is used to determine the number of factors, excessive sample size will result in a very large number of significant factors. The best safeguard against this possibility is careful experimental design in which the investigator knows in advance the approximate number of factors to expect by judicious selection of the variables.

Two rough rules of thumb serve as guides for appropriate sample size. These rules have no particular theoretical base, but do seem to offer handy limits which recognize that the chance for error covariation rises rapidly as the number of variables increases. The first limit, which can be taken as a minimum, is that there should be at least 10 subjects for every variable + 50. A more ideal sample size would be the square of the number of variables + 50 or $N = n^2 + 50$. This formula calls for a much more rapid increase in $N$ with increasing $n$ than does the first rule, but it seems justified in light of the increasing number of correlations in the matrix being analyzed. For a small study of 10 variables, both rules call for a sample of 150. When the number of variables is doubled to 20, the samples required by the two rules go to 250 and 450, respectively. In view of the fact that the number of off-diagonal correlations being estimated goes from 45 in the first instance to 190 in the second, the second rule would seem to be justified and preferable. (It is interesting that in a Monte Carlo study of the stability of canonical weights and loadings, Barcikowski and Stevens (1975) found that samples of approximately $N = n^2 + 50$ were necessary for the weights and loadings to show adequate stability.)

A second matter in the conduct of factor analyses is the care with which data are handled. This is a matter so fundamental to science that it really should not need repetition, but since there generally are more variables in multivariate research, additional care is necessary. The author is aware of more than one published investigation (names withheld to protect the guilty) in which the results reported are mathematically impossible in light of the data as published. Great care is necessary in the collection of the data to ensure that they are as accurate as possible and to prevent distortion from unwanted situational variation. Processing of the data requires equal care, and computer programs must be thoroughly checked with benchmark problems before and after the analysis to determine that they are working properly. Finally, transcription of the results for publication must be carefully done.

A final consideration of interest here is the choice of factoring method. We have discussed this issue elsewhere, but it is a topic that must be faced in conducting a study. The choice should be made on the basis of the purpose of the investigation and the use to be made of the results. In most studies the variables cannot be considered to have been sampled and therefore must be considered a population. This will generally rule out alpha factor analysis. If the subjects are also considered as a population rather than a sample (e.g., all patients in a particular hospital or all children in a school) then a least-squares factoring such as principal components with the Kaiser criterion may be used. If, however, the subjects are considered to be sampled from a larger definable population, then it is probably best to use a statistical factoring method such as minimum residuals or maximum likelihood. As we noted earlier, Harman's approach seems to differ little from Jörskog's, so either may be used. The only problem with these two methods is that they define the communalities for the investigator. If he has reason for using some other communality estimates, such as reliabilities, then he should do so and use a principal axis extraction procedure. Decisions about rotation must then be made, and we have considered these in detail elsewhere.

**Replication.** When a given factor analytic study is complete, the problem of how to tie it into a research program still can arise. We have already discussed the use of marker variables to tie a series of studies together. There are also occasions when we wish to replicate a factor structure in a new sample. This may take two forms, either a direct replication with another sample from the same population or a generalization study in which the goal is to compare factor structures in different populations.

These two objectives require different approaches, depending on the investigator's goals. The first response would probably be that Jöreskog's confirmatory factor analysis, or else some form of rotation to congruence or Procrustes procedure should be used. This line of attack will determine whether the factors found in one sample *can be found* in the other, and this is often an appropriate goal. However, it is also useful to perform an independent analysis of the new set of data, obtain a solution for those data that is most satisfying in terms of meaning and simple structure without regard to the first analysis, and then compare the two results. Thurstone (1947) argued that the "correct" simple structure solution would be compelling, unique, and would reappear in new samples that were independently analyzed. Therefore, independent factorings of similar sets of variables from similar samples should yield similar factors.

It should be remembered at this point that factor analysis is basically a descriptive technique rather than an hypothesis-testing one. The strict rules of the hypothesis-testing model do not apply, a fact that has caused criticism and problems. However, the search for an adequate descriptive frame allows some useful flexibility. For example, one approach to replicating factors might be to perform an independent analysis in the new sample and then apply a Procrustes rotation to determine maximum possible similarity. The amount of change needed to bring the independent solution into maximum agreement with the "target" solution might provide a useful index of the similarity of the two sets of factors. This line of research has not been pursued.

## INTERPRETATION OF FACTORS

After the mathematical aspects of factor analysis—extraction and rotation—have been completed it is common practice to give names to the factors. This sometimes questionable baptismal rite is the end result of the process of interpreting the factors. The investigator examines the factor loadings and generally interprets the factor as representing what is common to those variables that have high loadings on the factor. For example, a factor that has high loadings for several arithmetic tests would be interpreted as an arithmetic or computation factor. A more subtle example is the case of several tests with different content but all requiring rapid performance to obtain a high score. This might be labeled a speed or mental-quickness factor.

Factor interpretation is, in a sense, the investigator's reward for having performed the study. However, it is a task with many possible pitfalls, and it is unwise to place too much confidence in an interpretation of a factor that appears in a single study. A factor must reappear in a series of studies, and fail to appear in studies designed to exclude it, before much faith can be put in its existence and scientific value.

There are a couple of points an investigator must keep in mind when interpreting factors. One of these is that low loadings, those near zero, may be important in that they show what the factor does not contain. If, as in the second example, there are several arithmetic tests in the battery and only some of them require speed, it would be important to observe that tests of similar content that are nonspeeded load on a different factor from those which are speeded. In a battery of 12 tests, 3 speeded arithmetic, 3 nonspeeded arithmetic, 3 speeded verbal, and 3 nonspeeded verbal, one might even find an arithmetic factor with high loadings for the first six tests, a verbal factor with high loadings for the last 6 tests, a speed factor with high loadings for the speeded tests, and a power factor that had high loadings for the nonspeeded tests. Of course, this is an oversimplified example, but it is important to consider all the variables in interpreting each factor.

A second point to keep in mind is that three variables must have moderately high loadings on a factor before we can have confidence in its location and interpretation. This is a mathematical requirement that stems from our earlier discussion of the geometry of factor analysis. All too often attempts are made to name factors associated with single variables or pairs of variables. This should be avoided by a careful selection of the variables to be included in a study of a well-known domain or by performing a series of studies in a new domain before publication.

Another issue that arises at this stage is the problem of how large a loading must be before the variable warrants interpretation as an important contributor to the factor. There has been some controversy over this issue, with no satisfactory resolution. One view says that the decision should be a function of sample size. When the sample size is large, the standard errors of the factor loadings should decrease, and it should therefore be permissible to include relatively small loadings for interpretation, because they are unlikely to be chance deviations from zero. This position seems reasonable, but has three drawbacks: the estimation of sampling errors is rather complex and requires some additional computations, the sampling distribution of factor loadings is not known precisely, and very small loadings would bear interpretation when very large samples are used.

The most popular method of determining which factor loadings to consider for interpretation is to set up an arbitrary cut-off value (generally .3, .4, or .5) and retain for interpretation only those loadings that exceed the selected value, calling the rest zero. This approach is similar in logic to the Kaiser criterion and would appear to offer a useful solution, in that each loading reflects at least a useful portion of variance. However, a refinement may be in order. The Kaiser criterion was developed under the assumption that an unreduced matrix was being analyzed, and its use ensures that each factor retained will account for at least $1/n$ the of the variance in the matrix. When a reduced correlation matrix with estimated communalities in the main diagonal is used, the same logic would require that the minimum eigenvalue be equal to the average communality. It would seem useful also to apply this approach to the decision about the smallest factor loading to be retained. If a full correlation matrix has been analyzed, a reasonable lower limit for a factor loading might be .5, meaning that 25% of the variable's variance is associated with the factor. If, on the other hand, communality estimates have been used in the main diagonal, the square root of 25% of the average communality would seem to offer a good minimum value. However, whatever decision rule is used, the whole pattern of high and zero loadings must be considered in arriving at an interpretation.

Regardless of what choices are made about the number of factors, minimum factor loadings to be considered nonzero, communality estimates, rotation, and so forth, there is one practice in factor interpretation that appears from time to time, even in the work of the most eminent factor analysts, which should be strictly avoided. This is the practice of overlooking the loading of a variable on one factor in favor of a smaller loading on another. If a loading of .4 on one factor is considered real, then a loading of .5 on another factor, even though it is contrary to expectation or to one's pet theory, must also be considered real and given its due. To do less reflects badly on the investigator and his work.

## SUMMARY

In this chapter we have considered several issues related to the use of factor analysis. The problem of cross-validation in factor analysis has been discussed, with special attention to confirmatory studies. The uses of factor scores and a method for their estimation were then described. It was noted that the term factor score has a precise meaning and is often misused to refer to a more general composite of scores on groups of variables.

Next, some considerations in the application of factor methodology were presented. The appropriate place for factor analysis is in programmatic research rather than in one-shot studies. Careful planning and selection of both variables and subjects is a necessary but often overlooked requirement of a good factor study. Marker variables are of significant value and adequate sample size is a must. A suggested rule of thumb is that $N$ should be approximately equal to $n^2 + 50$. The need for great care in the handling of data was also emphasized, and the need for various replication strategies was discussed. Finally, some issues in the interpretation of factors was presented. It is necessary to consider all the variables in interpreting each factor. Customarily, those loadings that exceed a specified percentage of the average communality may be considered to be nonzero, and all others are considered zero. It is inappropriate to overlook one loading in favor of a smaller one that fits one's theory.

# REFERENCES

Ahmavaara, Y., Transformation analysis of factorial data. *Annals of the Finnish Academy of Science*. Helsinki, 1954.

Armstrong, J. S. and Soelberg, P., On interpretation of factor analysis. *Psychological Bulletin,* 1968, *70*, 361–364.

Barcikowski, R. S. and Stevens, J. P., A monte carlo study of the stability of canonical correlations, cannonical weights, and canonical variate-variable correlations. *Multivariate Behavioral Research*, 1975, *10*, 353–364.

Bartlett, M. S., The statistical significance of canonical correlations. *Biometrika*, 1941, *32*, 29–38.

———, Multivariate analysis. *Supplement to the Journal of the Royal Statistical Society*, 1947, *9*, 176–197.

———, Internal and external factor analysis. *British Journal of Psychology (Statistical Section)*. 1948, *1*, 73–81.

———, Tests of significance in factor analysis. *British Journal of Psychology (Statistical Section)*, 1950, *3*, 77–85.

Bolton, B., et al., *Annotated Bibliography: Factor Analytic Studies 1941–1970* (Vols. I–IV). Arkansas Rehabilitation Research and Training Center, University of Arkansas, 1973.

Borgen, F. H. and Weiss, D. J., Cluster analysis and counseling research. *Journal of Counseling Psychology*, 1971, *18*, 583–591.

Burt, C. L., Tests of significance in factor analysis. *British Journal of Psychology (Statistical Section)*, 1952, *5*, 109–133.

——, Appropriate uses of factor analysis and analysis of variance. In R. B. Cattell (Ed.), *Handbook of Multivariate Experimental Psychology*. Chicago: Rand-McNally, 1966, pp. 267–287.

Butler, J. M., Descriptive factor analysis. *Multivariate Behavioral Research*, 1968, *3*, 355–370.

Butler, J. M. and Hook, L. H., Multiple factor analysis in terms of weighted regression. *Educational and Psychological Measurement*, 1966, *26*, 545–564.

Carroll, J. B., An analytic solution for approximating simple structure in factor analysis. *Psychometrika*, 1953, *18*, 23–38.

Carroll, J. D. and Chang, J. J., Analysis of individual differences in multidimensional scaling via an N-way generalization of "Eckert-Young" decomposition. *Psychometrika*, 1970, *35*, 283–319.

Cattell, R. B., Extracting the correct number of factors in factor analysis. *Educational and Psychological Measurement*, 1958, *18*, 791–838.

——, The scree test for the number of factors. *Multivariate Behavioral Research*, 1966a, *1*, 245–276.

——, The meaning and strategic use of factor analysis. In R. B. Cattell (Ed.), *Handbook of Multivariate Experimental Psychology*. Chicago: Rand McNally, 1966b, pp. 174–243.

—— and Baggaley, A. R., The salient variable similarity index for factor matching. *British Journal of Statistical Psychology*, 1960, *13*, 33–46.

—— and Dickman, K., A dynamic model of physical influence demonstrating the necessity of oblique simple structure. *Psychological Bulletin*, 1962, *59*, 389–400.

—— and Jaspers, J., A general plasmode (No. 30-10-5-2) for factor analytic exercises and research. *Multivariate Behavioral Research Monographs No. 67-3*, 1967.

—— and Muerle, J. L., The "maxplane" program for factor rotation to oblique simple structure. *Educational and Psychological Measurement*, 1960, *20*, 569–590.

Cliff, N. and Hamburger, C. D., The study of sampling errors in factor analysis by means of artificial experiments. *Psychological Bulletin*, 1967, *68*, 430–445.

Cooley, W. W. and Lohnes, P. R., *Multivariate Procedures for the Behavioral Sciences*. New York: Wiley, 1962.

—— and Lohnes, P. R., *Multivariate Data Analysis*. New York: Wiley, 1971.

Coombs, Clyde, H., *A Theory of Data*. New York: Wiley, 1964.

Cronbach, L. J., Coefficient alpha and the internal structure of tests. *Psychometrika*, 1951, *16*, 297–334.

—— and Gleser, G. C., Assessing similarity between profiles. *Psychological Bulletin*, 1953, *50*, 456–473.

—— and Gleser, G. C., *Psychological Tests and Personnel Decisions* (2nd ed.). Urbana: University of Illinois Press, 1965.

Cronbach, L. J., Gleser, G. C., Nanda, H. and Rojaratnam, N., *The Dependability of Behavioral Measurements: Theory of Generalizability for Scores and Profiles*. New York; Wiley, 1972.

Cureton, E. E., Validity, reliability and baloney. *Educational and Psychological Measurement*, 1950, *10*, 94–96.

Draper, N. R. and Smith, H., *Applied Regression Analysis*. New York: Wiley, 1966.

Dunnette, M. D., A note on *the* criterion. *Journal of Applied Psychology*, 1963, *47*, 251–254.

Eber, H. W., Toward oblique simple structure: Maxplane. *Multivariate Behavioral Research*, 1966, *1*, 112–125.

——, Maxplane meets Thurstone's "factorially invariant" box problem. *Multivariate Behavioral Research*, 1968, *3*, 249–254.

Ezekiel, M. and Fox, K. A., *Methods of Correlation and Regression Analysis* (3rd ed.). New York: Wiley, 1957.

Flanagan, J. C., et al., *Design for a Study of American Youth*. Boston: Houghton Mifflin Co., 1962.

Fleishman, E. A. and Hempel, W. E., Jr., A factor analysis of dexterity tests. *Personnel Psychology*, 1954, *7*, 15–32.

Fruchter, B., *Introduction to Factor Analysis*. New York: Van Nostrand, 1954.

Gough, H. G., *California Psychological Inventory*. Palo Alto, Calif.: Consulting Psychologists Press, 1956.

Guilford, J. P., *Psychometric Methods* (2nd ed.). New York: McGraw-Hill, 1954.

—— and Fruchter, B., *Fundamental Statistics in Psychology and Education* (5th ed.). New York: McGraw-Hill, 1973.

Guttman, L., Some necessary conditions for common-factor analysis. *Psychometrika*, 1954, *19*, 149–161.

———, "Best possible" systematic estimates of communalities. *Psychometrika*, 1956, *21*, 273–285.

Harman, H. H., *Modern Factor Analysis* (2nd ed.). Chicago: University of Chicago Press, 1967.

——— and Fukuda, Y., Resolution of the Heywood case in the minres solution. *Psychometrika*, 1966, *31*, 563–571.

Hays, W. L., *Statistics for Psychologists*. New York: Holt, Rinehart and Winston, 1963.

Horn, J. L., A rationale and test for the number of factors in factor analysis. *Psychometrika*, 1965, *30*, 179–186.

Horst, P., Relations among *m* sets of measures. *Psychometrika*, 1961, *26*, 129–149. (reprint)

———, *Matrix Algebra for Social Scientists*. New York: Holt, Rinehart and Winston, 1963.

Hotelling, H., Analysis of a complex of statistical variables into principal components. *Journal of Educational Psychology*, 1933, *24*, 417–441, 498–520.

———, The most predictable criterion. *Journal of Educational Psychology*, 1935, *26*, 139–142.

———, Relations between two sets of variates. *Biometrika*, 1936, *28*, 321–377.

Humphreys, L. G., Number of cases and number of factors: An example where N is very large. *Educational and Psychological Measurement*, 1964, *24*, 457–466.

Jones, J. K., *The Multivariate Statistical Analyzer*. Cambridge, Mass.: Harvard Cooperative Society, 1964.

Jöreskog, K. G., Some contributions to maximum likelihood factor analysis. *Psychometrika*, 1967, *32*, 443–482.

———, A general approach to confirmatory maximum likelihood factor analysis. *Psychometrika*, 1969, *34*, 183–202.

Kaiser, H. F., The varimax criterion for analytic rotation in factor analysis. *Psychometrika*, 1958, *23*, 187–200.

———, The application of electronic computers to factor analysis. *Educational and Psychological Measurement*, 1960, *20*, 141–151.

———, A note on Guttman's lower bound for the number of common factors. *British Journal of Statistical Psychology*, 1961, *14*, 1–2.

Kaiser, H. F. and Caffrey, J., Alpha factor analysis. *Psychometrika*, 1965, *30*, 1–14.

Kendall, M. G., Hiawatha designs an experiment. *American Statistician*, 1959. Reprinted in Jackson, D. N. and Messick, S. (Eds.) *Problems in Human Assessment*. New York: McGraw-Hill, 1967.

Lawley, D. N., The estimation of factor loadings by the method of maximum likelihood. *Proceedings of the Royal Society of Edinburgh*, 1940, *60*, 64–82.

———, The application of the maximum likelihood method to factor analysis. *British Journal of Psychology*, 1943, *33*, 172–175.

———, Tests of significance for the latent roots of covariance and correlation matrices. *Biometrika*, 1956, *43*, 128–136.

Linn, Robert, L., A monte carlo approach to the number of factors problem. *Psychometrika*, 1968, *33*, 37–71.

Mahalanobis, P. C., On generalized distance in statistics. *Proceedings National Institute of Science, India*, 1936, *12*, 49–58.

Maxwell, A. E., Statistical methods in factor analysis. *Psychological Bulletin*, 1959, *56*, 228–235.

McDonald, R. P., A note on Rippe's test of significance in common factor analysis. *Psychometrika*, 1975, *40*, 117–119.

McNemar, Q., On the number of factors. *Psychometrika*, 1942, *7*, 9–18.

———, *Psychological Statistics* (4th ed.). New York: Wiley, 1969.

Meehl, P. E. and Rosen, A., Antecedent probability and the efficiency of psychometric signs, patterns, or cutting scores. *Psychological Bulletin*, 1955, *52*, 194–216.

Morrison, D. F., *Multivariate Statistical Methods*. New York: McGraw-Hill, 1967.

Mosier, C. I., Problems and designs of cross-validation. *Educational and Psychological Measurement*, 1951, *11*, 5–11.

Mulaik, S. A., Inferring the communality of a variable in a universe of variables. *Psychological Bulletin*, 1966, *66*, 119–124.

Neuhaus, J. O. and Wrigley, C., The quartimax method: An analytic approach to orthogonal simple structure. *British Journal of Statistical Psychology*, 1954, *7*, 81–91.

Norman, W. T., Double-split cross-validation: An extension of Mosier's design, two undesirable alternatives, and some enigmatic results. *Journal of Applied Psychology*, 1965, *49*, 348–357.

Nunnally, J. C., *Introduction to Psychological Measurement*. New York: McGraw-Hill, 1970.

Pearson, K., *Tables for Statisticians and Biometricians, Part II*. Cambridge: Cambridge University Press, 1931.

Rao, C. R., Estimation and tests of significance in factor analysis. *Psychemetrika*, 1955, *20*, 93–111.

Reyburn, H. A. and Raath, M. J., Simple structure: A critical examination. *British Journal of Psychology* (*Statistical Section*), 1949, *2*, 125–133.

Rippe, D. D., Application of a large sampling criterion to some sampling problems in factor analysis. *Psychometrika*, 1953, *18*, 191–205.

Roff, M., Some properties of the communality in multiple factor theory. *Psychometrika*, 1935, *1*, 1–6.

Saunders, D. R., An analytic method for rotation to orthogonal simple structure. *Research Bulletin, RB 53-10*. Princeton, N. J.: Educational Testing Service, 1953.

Sheppard, R. N., Representation of structure in similarity data: Problems and prospects. *Psychometrika*, 1974, *39*, 373–421.

Siegel, S., *Nonparametric Statistics for the Behavioral Sciences*. New York: McGraw-Hill, 1956.

Sokal, R. R., A comparison of five tests for completeness of factor extraction. *Transactions of the Kansas Academy of Science*, 1959, *62*, 141–152.

Stevens, S. S., On the theory of scales of measurement. *Science*, 1946, *103*, 677–680.

Stewart, D. and Love, W., A general canonical correlation index. *Psychological Bulletin*, 1968, *70*, 160–163.

Taylor, H. C. and Russell, J. T., The relationship of validity coefficients to the practical effectiveness of tests in selection. *Journal of Applied Psychology*, 1939, *23*, 565–578.

Thomson, G. H., Hotelling's method modified to give Spearman's g. *Journal of Educational Psychology*, 1934, *25*, 366–374.

Thorndike, R. M., Simple Structure: A revision and an objective criterion. *Psychometric Society*, April, 1971. (mimeograph).

———, Canonical analysis and predictor selection. *Multivariate Behavioral Research*, 1977, *12*, 75–87.

——— and Weiss, D. J., An empirical investigation of stepwise canonical correlation. *41st Annual Meeting of the Midwestern Psychological Association, May, 1969*. (mimeograph).

———— and Weiss, D. J., A study of the stability of canonical correlations and canonical components. *Educational and Psychological Measurement*, 1973, *33*, 123–134.

Thurstone, L. L., Multiple factor analysis. *Psychological Review*, 1931, *38*, 406–427.

————, *The Theory of Multiple Factors*. Ann Arbor, Mich.: Edwards Bros., 1933.

————, Primary mental abilities. *Psychometric Monographs*, No. 1, 1938.

————, *Multiple Factor Analysis*. Chicago: University of Chicago Press, 1947.

Tobias, S. and Carlson, J. E., Brief report: Bartlett's test of sphericity and chance findings in factor analysis. *Multivariate Behavioral Research*, 1969, *4*, 375–377.

Torgeson, W. S., *Theory and Methods of Scaling*. New York: Wiley, 1958.

Tryon, R. C., General dimensions of individual differences: Cluster analysis vs. multiple factor analysis. *Educational and Psychological Measurement*, 1958, *18*, 477–475.

———— and Bailey, D. E., *Cluster Analysis*. New York: McGraw-Hill, 1970.

Veldman, D. J., *FORTRAN Programming for the Behavioral Sciences*. New York: Holt, Rinehart and Winston, 1967.

Ward, J. H., Hierarchical grouping to optimize an objective function. *Journal of the American Statistical Association*, 1963, *58*, 236–244.

Wrigley, C., The distinction between common and specific variance in factor theory. *British Journal of Statistical Psychology*. 1957, *10*, 81–98.

————, Objectivity in factor analysis. *Educational and Psychological Measurement*, 1958, *18*, 463–476.

————, The effect upon the communality of changing the estimate of the number of factors. *British Journal of Statistical Psychology*, 1959, *12*, 34–54.

# INDEX

**This book was set in Times Roman
on the Monophoto by Service Type Inc., Lancaster, Pennsylvania.
It was printed and bound by
Capital City Press, Montpelier, Vermont.
Design by Sidney Solomon.**

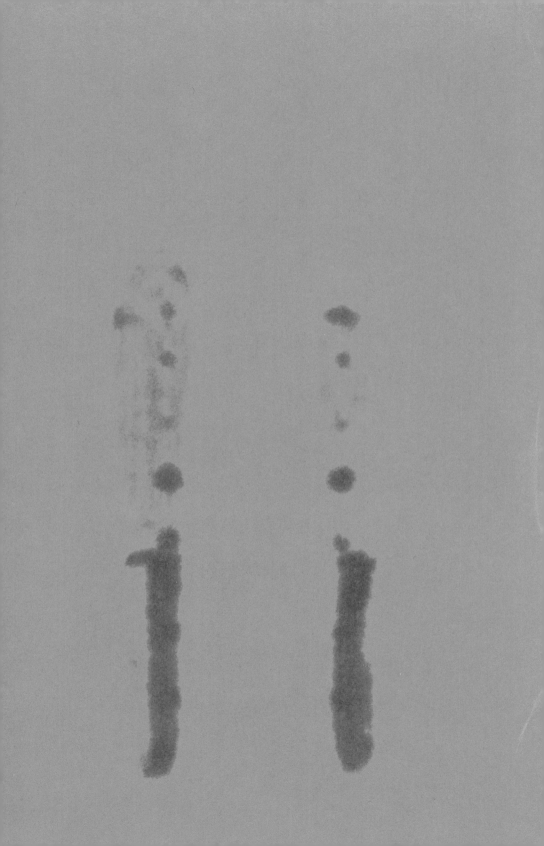

BETQ